The Shape of Consciousness

How Finite Minds Filter a World Into Being

Scott E. Johnson

Inner Exploration Labs

Published by Inner Exploration Labs

Paperback ISBN:	979-8-9965451-0-0
Hardcover ISBN:	979-8-9965451-2-4
EPUB ISBN:	979-8-9965451-1-7

First Edition: 2026

Publisher site: innerexplore.com

Companion visualizations: innerattractors.com

In memory of my Mother,
Rosemary Livingston,
who saved me from the world
by giving me free rein to explore it.

Contents

Preface

Almost everything is missing from human consciousness. At any moment, more of the world is reaching you than any mind could possibly hold: light outside the visible range, pressure variations in the air, chemical traces your senses barely register, bodily signals humming beneath awareness, past experiences stored in muscle and habit, and social meanings flickering at the edge of attention. You are not experiencing reality as a whole. You are experiencing what your mind and body have selected, weighted, stabilized, and made livable.

This does not diminish consciousness; it is the condition that makes it possible. A finite being must filter the world, allowing some things through and keeping others out. What we call ordinary experience is already the result of an enormous act of selective filtering.

But this filter is never entirely fixed; it constantly reshapes itself in response to what it encounters. Sometimes the mind narrows until only one path seems available. Sometimes it opens until the self's usual boundaries become porous. Grief reorganizes time; fear narrows the available world to a single demand; love opens the self to contain another life. Music, terror, illness, psychedelics, or sudden insight can change not merely what we think but the very shape of the landscape in which thinking occurs.

I learned this early, and not only from books. I came to know that consciousness has thresholds, that the mind can close, flood, split, return, and reorganize. Later, I would find this same structure in psychology, in addiction and recovery, in contemplative traditions, and eventually in the strange new relation between human beings and silicon systems.

My professional life, however, has not been focused on the philosophy of mind. I have spent my career as an Earth scientist, studying systems that are too large, too slow, or too hidden to be grasped all at once: rocks that preserve histories no one witnessed, landscapes shaped by forces acting across deep time, and climate systems whose visible patterns emerge from interactions distributed across the whole planet. Studying Earth systems trains a certain kind of attention. You learn to infer processes from trace evidence, history from structure, and instability from small signs in seemingly stable behavior. You learn to think in terms of basins, thresholds, feedbacks, accumulated stresses, sudden reorganizations, and regimes that can persist for long periods before changing state.

That training shaped how I came to think about consciousness, but it was not the only catalyst. Long before I mapped physical systems, early experiences with psychedelics had already shown me that the mind's boundaries are not absolute: the filter can open and the self can radically reorganize. Earth science simply gave me the formal language for what those perturbations had first revealed. Because of both, I came to see the mind as a landscape: patterned, historical, constrained, and dynamic. Some mental states are easy to enter and hard to leave. Some thresholds are crossed slowly at one time and suddenly at another. Some patterns consolidate until they feel like identity itself, whereas others erode, loosen, or reorganize under pressure.

This book is therefore not a departure from my professional field so much as an extension of a habit learned there: read the visible pattern, infer the hidden dynamics, and ask what kind of system could have produced this shape. The project began when I was trying to formalize one piece of that dynamical pattern: the way people change. I was working on a model of psychological transition called

Engagement, Openness, and Consolidation (EOC), a three-variable system for thinking about how people approach change, loosen rigid habits, and consolidate new ones.

But the geometry of that psychological model began to show me more than I had anticipated. It suggested that inner life is not a sequence of static snapshots; it has motion. It is drawn toward persistent patterns that dynamical systems theory calls *attractors.* It has basins where we get stuck, thresholds we cross, and returns to the familiar. Conscious life has a geometry. In this book, I call that geometry the experiential landscape.

The central claim that follows is simple to state, even if it takes the rest of the book to describe: consciousness is what it is like, from the inside, to be a creature that selects from the world, evaluates its own selecting, and cares about what the selection does to the life it is trying to maintain. What comes next is the case for that claim and a map of where it leads.

Acknowledgments

I thank the friends and colleagues who tolerated my nine-month obsession with this book. Several AI systems assisted with background research, coding for the figures, critical review, and book production.

Part I

The Architecture of Consciousness

Chapter 1

Why This Book, Why Now

1.1 Why Consciousness Matters

Before consciousness became a research topic in philosophy, it was already the medium of ordinary human experience. Pain hurts. Grief can alter the texture of time. Music can change the atmosphere of a room. Wonder opens experience outward. Love makes another life matter from the inside. Consciousness is the medium in which anything matters to us at all. A world without consciousness could still contain structure, motion, and information, but it would lack any felt register of events: no joy associated with them, no terror, no beauty, no humiliation, and no relief.

That is why the problem has spread into multiple disciplines. Philosophy, physics, neuroscience, and psychology all approach it from different directions. Philosophers confront its resistance to easy reduction; physicists ask whether anything in fundamental theory leaves room for it; neuroscientists observe how experience shifts when brains shift; psychologists find felt life entangled with attention, memory, emotion, and selfhood. Ordinary people wrestle with it too because nothing in their lives feels more immediate or more mysterious than the bare fact that the world shows up at all.

The stakes are ethical as well as scientific. Any serious account of consciousness spills quickly into ethics. If there is something it is like to be a creature, then that creature can be benefited or harmed in a deeper sense than mere mechanical disruption. A conscious life can go badly from the inside. That is true of human beings and is increasingly clear for many other animals as well. And it may, on grounds the book will lay out, turn out to be true of artificial systems we are now building. So a theory of consciousness bears on ethics as much as on explanation: on suffering, responsibility, personhood, and the boundaries of moral regard.

1.2 Why the Question Has Become Newly Urgent

For a long time, consciousness could be treated as a problem with no real deadline. Philosophers could puzzle over zombies, neuroscientists could map correlates, clinicians could study altered states, and the wider culture could proceed on the broad assumption that conscious beings were the familiar ones: humans, perhaps some animals, and nothing made of code and silicon. That assumption no longer feels as settled as it once did.

Whether a system built on another substrate could become conscious is still an open question. So is the question of whether the systems we are currently building show any of the organizational precursors that would matter. What has changed is proximity. We now live alongside systems that reason in language, sustain context over long stretches, collaborate across extended projects, revise their own outputs, and sometimes behave as if they are tracking the constraints of the situation in which they are being used or evaluated. None of that proves consciousness, but it does make the question harder to postpone. If consciousness is a matter of organization rather than substance, the real issue is what kind of organization counts, what evidence bears on the judgment, and how we recognize when an old categorical boundary has lost its original justification and persists out of habit.

AI matters here for two reasons: it makes the question of consciousness newly urgent, and it gives the framework developed here a hard contemporary test. It turns a philosophical puzzle into an empirical and ethical judgment we can no longer avoid.

1.3 What This Book Claims

The argument presented here is that consciousness is not a second substance added to physical processes, nor is it an illusion produced by machinery with no inner life. It is what it is like, from the inside, to be a system that filters the world, models its own filtering, and has something at stake in the result of that filtering. You will see this referred to as *recursive stake-laden filtering* or similar language throughout.

That claim will be unpacked across Part I, but the underlying idea is fairly intuitive. Every bounded system confronts more structure than it can process or use. To persist, it has to select. It must couple to some features of the world while ignoring the rest. That selective coupling is what I call filtering, and most of it is unconscious. Membranes filter, thermostats filter, and nervous systems filter; but only certain forms of filtering become conscious. The threshold presented here is reached when filtering becomes deep enough to include an ongoing model of the system's own filtering and when the outcomes of that filtering bear on the system's continued integrity. At that point, the system is not only regulating its interaction with the world, but it is also tracking itself as something for which things can go better or worse.

If that is right, the so-called hard problem of consciousness moves. The old question was how non-experiential stuff could manufacture experience as an extra ingredient. The new question is which forms of recursive, stake-laden filtering are identical with experience. The work of consciousness research then becomes mapping how those forms vary, how they fragment, and under what conditions they might arise in different substrates.

The framework's answer to qualia, the philosophers' term for what experience is like from the inside, follows from the same identity claim. The felt character of experience is the lived form of motion through the system's own experiential landscape. The landscape is the terrain on which motion can occur, and the motion itself is what defines lived experience. That is why the EOC landscape appears later in the book, with its formal backbone summarized in the EOC Primer. Its value is not that it captures the whole psyche in three variables, but that it makes the central claim visible. It lets the reader observe residence, return, crossing, instability, and regime change as motion through a landscape. For a theory that treats qualia as trajectories, seeing those trajectories in motion matters.

1.4 How the Book Is Organized

The first part of the book builds the conceptual architecture. It starts by revisiting the hard problem, asking whether its deepest difficulty is actually generated by an unnecessary partition between mechanism and experience. It then argues that filtering comes first for any bounded system, that consciousness is a special case of filtering rather than something added from outside, that stakes are necessary for inner life, and that qualia can be understood as trajectories through a structured attractor landscape rather than as ineffable extras.

Part I asks a good deal of the reader, so before the formal turn to Part II, I pause for an Interlude in Chapter 10. That chapter restates the argument in plainer language and gives the reader a place to catch their breath before the visual dynamical examples arrive.

Part II then stress-tests the framework. Chapter 11, "The Shape of Qualia," uses the EOC landscape as the minimum visualizable case of qualia-as-trajectory. It shows what the identity thesis looks like when phenomenal character is understood as motion through a landscape. Chapter 12 then uses that result to revisit archetypes, treating the Shadow, Hero, Trickster, Great Mother, and Self as recurring landscape features, not inherited pictures.

The chapters that follow carry the argument across deep time, altered states, addiction and recovery, and the carbon-silicon conversation. The final chapter returns to the work still ahead, the uncertainties that remain, and the question still waiting at the edge of the framework: not just how finite filters become conscious, but what a universe must be like to produce filters that can.

1.5 Where This Book Sits Among Other Theories

The account of consciousness presented here arises from several long-running conversations. Some theories emphasize global availability: consciousness arises when information becomes available across the whole system, so that what one part takes in, the rest can act on. Some emphasize self-modeling: conscious states are the ones a system represents as its own. Some emphasize integration: what matters is whether the system as a whole has features that no division of its parts can capture. Some emphasize prediction and regulation: consciousness reflects how a system anticipates what is coming and reduces its uncertainty about the world. Others emphasize embodiment and sense-making: consciousness belongs to living systems that maintain themselves in active relation to a world.

This book draws selectively from several traditions: filter theories, self-model and strange-loop accounts, embodied and predictive approaches, geometric theories of consciousness, and neurodynamical work on complex systems. Four commitments hold the argument together: conscious systems are selective, choosing what to attend to from a much larger flow of possibilities; recursive, modeling themselves as systems that model; stake-laden, processing outcomes that shape what comes next; and dynamical, moving through experiential landscapes with shape and direction over time. Where the book departs from its neighbors is in tying those commitments to a direct identity claim. Recursive, stake-laden filtering is not merely *correlated* with experience; rather, it *is* the experience viewed from the inside. Readers who want the fuller comparative map will find it in the Extended Historical Perspective Appendix.

The chapters that follow begin where the lived problem actually begins. We are conscious creatures in a world too large to be taken in whole, and the first fact of our situation is that finite beings must select.

Chapter 2

Reframing the Hard Problem Through Filtering

2.1 Reframing the Hard Problem

Long before consciousness became a formal problem in philosophy, it was part of ordinary life. Coffee smells like something; pain is not merely information but real hurt; music can make the whole room seem altered. Theory came later and tried to account for what was already obvious from the inside.

In 1995, David Chalmers (Chalmers, 1995b) gave the problem of consciousness its now-standard formulation. He separated the questions that neuroscience and cognitive science already knew how to approach, in their own terms, from the one question that seemed to resist those terms even after all the mechanisms were in place. Neural firing, information integration, attention, report, discrimination, control: these are what Chalmers called the easy problems, not because they are trivial, but because they yield in principle to the ordinary methods of science. The remaining question is of another kind: why should any of this processing be accompanied by felt experience at all? Why should there be something it is like to smell coffee, see red, or sit with grief?

That framing clarified the terrain, made it harder to mistake a neural correlate for an explanation of experience itself, and gathered into one memorable distinction a set of earlier insights associated with Nagel (1974), Jackson (1982), Levine (1983), and others who each saw part of the same difficulty. The modern interdisciplinary study of consciousness owes a great deal to that formulation.

Still, the hard problem, as it is usually posed, begins by splitting the world in two. On one side are physical processes, and on the other, experience. Once that division is in place, philosophy and science are left trying to construct a bridge between them. The bridge may be described as correlation, emergence, supervenience, constitution, or identity. But the problem is already there, built into the initial separation. If process and experience are taken to be fundamentally different kinds of things, any bridge between them is bound to look mysterious.

The question this book asks is whether that partition is itself responsible for generating much of the difficulty. The distinction between the third-person description and first-person presence is real enough, but there may not be two separate kinds of fact here waiting to be connected. There may instead be a single organized process, describable from the outside in dynamical terms and lived from the inside as experience.

A whirlpool is what water is doing when organized in a particular way, not a second substance added to water. Change the organization and the whirlpool disappears, not because an extra thing has departed, but because the flow has changed. The identity thesis proposed in this book treats consciousness in the same spirit. Experience is not an ornament laid on top of a system that filters, models its own filtering, and has stakes in the result. It *is* that organization under first-person description.

This is a substantive claim. If it is right, the central scientific question changes. We stop asking how matter produces felt experience as an extra ingredient and ask what kind of organization makes a finite system present to itself. That question is still difficult, but it is exposed to the ordinary pressures of scientific work: specify the organization, compare cases, measure what can be measured, and let the theory fail where it should fail.

There is another reason to think the inherited framing may not be final. Experience is not fixed; instead, it deepens, contracts, fragments, quiets, and sometimes disappears. Anesthesia can flatten it. Psychedelics reorganize it. Meditation may calm or clarify it. Dreams redistribute it. Severe affective states can narrow it until only a small band of possibilities remains visible. None of these changes, by themselves, establish an identity thesis. They do, however, suggest that felt character is tied more tightly to dynamical organization than we usually notice.

The claim developed in the chapters ahead is that experience need not be treated as a separate ontological layer. The mystery is real, but we may have been posing it in a way that makes it harder than it needs to be. The task, then, is to understand a world organized deeply enough to appear from within.

If the partition between process and experience has been making the hard problem harder than it needs to be, the next question is where the alternative picture should begin. It does not begin with experience, because the felt dimension is what we are trying to explain. It does not begin with brains in particular, because a brain-first account can smuggle in the conclusion that consciousness requires one specific substrate before the argument has been made. It begins with something more general: the predicament of any finite system facing a world that exceeds it. The system cannot take in everything, so it must select.

2.2 Why Filtering Comes First

Every finite system is bounded: its state capacity is limited, as are its time, energy, and access to the structure around it. No such system can couple equally to everything in its environment; if it tried, it would be overwhelmed. To persist, it has to be selective.

Think for a moment about what is physically arriving at your body right now: electromagnetic radiation across a wide spectrum; acoustic vibrations layered into complex waveforms; chemical traces from wood, fabric, skin, food, and air; subtle gradients of temperature, humidity, and pressure; cosmic radiation, gravitational influences, and mechanical stress. None of this is hidden, but it is just too much. Physical reality reaches us at a bandwidth that exceeds the processing capacity of any finite system by many orders of magnitude.

The first task of a bounded system is triage. Some features of the world need to be taken up, stabilized, weighted, and put to use; most need to be ignored. In its broadest sense, a filter is any organized system that couples selectively and repeatably with some features of the world rather than with others. I use the word *filter* at several levels in this book, so one distinction is important. Selective coupling is not yet adaptation. Adaptation is not yet self-modeling. Self-modeling is not yet conscious life. The family resemblance is real, but each threshold adds new structure: selective coupling, adaptive coupling with history, self-modeling coupling, and finally, stake-laden self-modeling.

Membranes filter molecules; immune systems filter pathogens; retinas filter light; attention filters experience; the brain filters a flood of uncertain signals into a workable picture of the world. Across very different systems and very different scales, the same necessity keeps reappearing: more arrives than can be taken in, and only what appears to be usable gets through.

That is why the language of filtering helps. Older versions of the idea sometimes drift toward a strong transmission picture, as though the brain or mind were simply reducing a cosmic consciousness already present in full. This book does not need to settle that larger question. In our present finite condition, the signal can be understood simply as the overwhelming fullness of physical reality as it reaches any bounded system. Conscious experience, as we live it, depends on filtering being organized deeply enough. What, if anything, lies beyond that finite condition is another question.

Figure 2.1 shows that asymmetry in its simplest form: a broad, unruly signal on one side, a bounded filter in the middle, and a small livable world on the other.

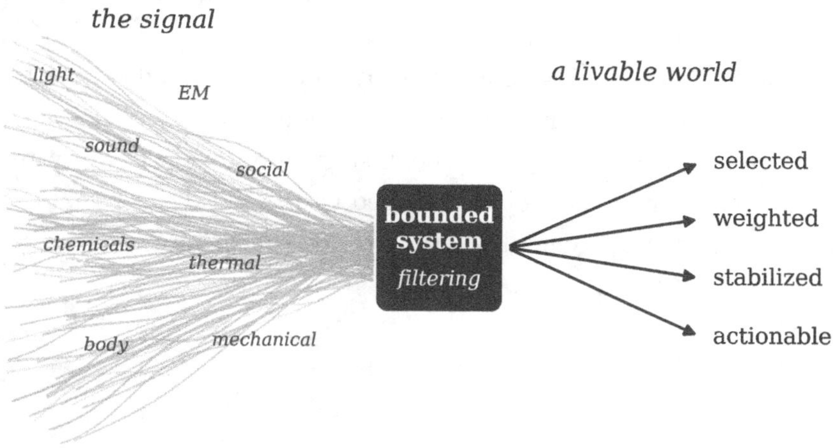

Figure 2.1. Filtering as the first condition of finite life. Physical reality reaches any bounded system at a rate that exceeds its processing capacity by many orders of magnitude. Light, sound, chemical traces, thermal gradients, mechanical pressure, social cues, body signals, and electromagnetic fields all arrive at once. No finite system can couple to the whole signal. The first task is therefore not knowledge in any grand sense but triage: select some features, ignore most, and stabilize a livable subset out of an overwhelming totality. The figure makes that asymmetry visible. The wide field on the left is the signal as it arrives; the narrow set of clean lines on the right is the world the system can actually live in.

Filtering alone is not consciousness. A thermostat registers a range and acts on it, but there is nothing it is like to be a thermostat. A calculator filters inputs into outputs without any inner life. Filtering is only where the story starts. The real question is what kind of filtering could ever become present to itself.

The answer proposed here adds two conditions. First, the filtering must become recursive: the system must carry an ongoing model of its own selection, regulation, and error. Second, the consequences of that filtering must bear on the system's continued integrity. Filtering becomes conscious when a bounded system filters in a way that includes itself, and when that self-model carries vulnerability within it.

That is why filtering comes first. It names the broadest, most unavoidable layer of the architecture. Consciousness is a special, highly organized case within a more general story about what finite systems must do if they are not to be swamped by the world. The next chapter turns from that general necessity to the specific organization that yields inner life.

Chapter 3

Consciousness as Recursive Filtering

Consciousness, in this view, begins when a filter becomes recursively self-modeling, modeling not just the world but its own modeling of the world, with something at stake. Imagine you are an experienced paddler moving down a river in a kayak. You do not try to overpower the current; you read it: eddies behind rocks, standing waves over shallow shelves, channels opening and closing. With practice, the body learns the pattern before it can fully name what it knows. The next movement feels right or wrong before it becomes a proposition. That feeling of rightness is what refined coupling with a structured flow is like from within.

The same basic logic applies inwardly. A scientist returning to a problem can feel tension gathering before the solution arrives. A meditator notices the mind tightening around one thought and loosening around another. An addict may feel the pull of an old habit long before action follows. In each case, the system is doing more than responding to conditions; it is also, to some degree, tracking its own response.

Recursive means the system carries a working sense of itself doing what it does. The paddler reading the river is, at some level, also sensing themselves reading the river: noticing when their attention is sharp and when it has slipped, when their body has the rhythm and when it has lost it, and when something is at stake and when it isn't. This self-tracking does not need to be in words. It can be wordless, fragmentary, or only partially present. What matters is that the system takes in the world and, alongside that, takes in itself as the thing doing the taking in.

Self-modeling means that the filter carries structured estimates of its own body, capacities, priorities, and likely futures. These estimates need not amount to a philosophically sophisticated self-concept. A creature can model itself well enough to regulate around injury, hunger, uncertainty, or social threat without narrating any of it. The self-model is practical before it becomes autobiographical.

Stakes enter the architecture when some trajectories bear differently on the system's continued integrity. The system can be harmed, depleted, frustrated, or stabilized; outcomes are not equivalent. From the outside, this looks like regulation around states that the system actively maintains or returns to. The next chapter will defend why, under the identity thesis, that same organization can be described from within as concern, valence, urgency, relief, dread, attachment, or care.

The hierarchy allows for gradation. A system need not leap from total darkness to full human-style consciousness in a single step. In this view, conscious depth is a continuum, not one magical threshold. A minimal system may register only crude valence and immediate viability, while richer systems may model bodily states along with counterfactual futures, social bonds, reputational standing, or normative commitments. The basic architecture remains the same, even as its depth and texture vary.

This graded approach becomes useful later when the book turns to AI. A threshold-only framework invites sterile arguments about whether a system is conscious in one decisive leap. A graded framework makes room for better questions: how much self-modeling is present, how stable it is across contexts, what kinds of stakes are operative, and how much regulation is integrated around a continuing model of itself, not local heuristics or policy imitation.

Table 3.1. The filtering hierarchy. F3 supplies self-modeling; F4 adds stakes and marks the minimal conscious architecture presented here.

Level	Name	What it adds	Illustrations
F1	Selective coupling	The system couples to some features of the signal and not to others.	Crystal filtering light; band-pass filter; membrane transport
F2	Adaptive coupling	The coupling pattern changes through history or learning.	Synaptic plasticity; immune adaptation; machine learning updates
F3	Self-modeling coupling	The dynamics include an operative model of the system's own filtering, integrated into regulation. Necessary for consciousness, but not sufficient by itself.	Body-state self-tracking; uncertainty estimation; self-model in action selection
F4	Stake-laden self-modeling	The self-model includes vulnerability and the maintenance of preferred or viable states. This is the minimal conscious architecture here.	Interoception; somatic markers; value protection; self-preserving regulation

Table 3.1 gives the hierarchy in compact form. Figure 3.1 makes the same point visually: each level contains the one below it, and F4 is not an extra glow added at the end but the point where self-modeling becomes stake-laden.

> *Filtering is necessary but not sufficient. Adaptive filtering is necessary but not sufficient. Self-modeling is necessary but not sufficient. Stake-laden self-modeling is the minimal architecture that this book identifies with consciousness.*

F4 is therefore a composite step. It marks the conjunction of F3 recursion with stakes integrated into self-organization. Chapter 4 develops the stakes hierarchy that fills out the F4 step, with viability stakes as the minimum required for consciousness, and relational and normative stakes elaborating conscious life beyond the threshold. The two hierarchies operate together: the F-hierarchy describes the architectural conditions for self-modeling, and the stakes hierarchy describes the depth of stakes integration that any conscious system carries.

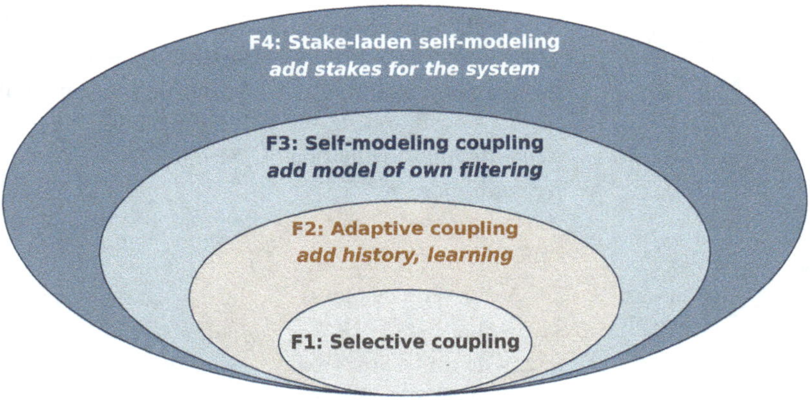

Figure 3.1. The filtering hierarchy. Selective coupling (F1) appears in matter itself, wherever a system couples to some features of the signal and not others. Adaptive coupling (F2) adds history and learning. Self-modeling coupling (F3) adds an ongoing internal model of the system's own filtering. Stake-laden self-modeling (F4), where the system's own continued integrity participates in the filtering, is the minimal architecture that this book identifies with consciousness. Each level includes and deepens the levels inside it; above the F4 threshold, the framework predicts a continuum of conscious depth.

From this, a minimal architecture follows. Three conditions appear necessary, and each is in principle measurable. First, the system must maintain a self-model that updates in real time rather than appearing only in retrospective report. Second, it must show recursive depth: in some form it must model its own modeling, not only the outside world. Third, that self-model must participate in regulation around states that matter for the system's continued integrity. Remove any one of these conditions, and the framework no longer predicts consciousness in the relevant sense.

The identity claim is stronger than the language of correlation. Much of the neuroscience of consciousness proceeds by locating processes that reliably accompany conscious states and calling them correlates. A correlate is a partner, not the thing itself; the relation between the partner and the thing is left unexplained. In the present view, recursive stake-laden filtering and experience are not partners. They are the same organized process under two complete descriptions: one from outside, one from within.

This is where the book parts company with views that allow a fully self-representing system to have no inner life. A system that recursively models its own attention, regulation, and stakes, with vulnerability inside the self-model, already meets the proposed identity conditions. Asking what further property might still be missing reinstates the gap between process and experience that the argument is built to close.

This is also why identity is preferred here to supervenience. Supervenience says that once the physical facts are fixed, the experiential facts are also fixed, but it still preserves two levels of fact: a base and something riding on it. The gap between them is exactly where the hard problem resides. Identity dissolves the gap, claiming one organized process available under two complete descriptions with no further fact to add.

The next chapter turns to the part most accounts treat as the end of the line: the felt character of experience itself.

CHAPTER 4

Stakes: Why Some Self-Models Feel Like Something

A system can model itself without having an inner life. It might keep an internal account of its own operations and remain indifferent, in any deep sense, to whether one future arrives rather than another. A diagnostic program running on a server may track the machine's operation and its own performance, but the integrity at stake is the server's uptime or the user's goal, not the diagnostic program as an organized subject. Such a system might be clever, responsive, even richly representational, and still miss the deeper structure this book treats as necessary for consciousness. A self-model becomes an inner life only when some trajectories count for more than others in the maintenance of the system's integrity.

Stakes name the organizational fact that certain states are preferentially maintained, that departures from them call forth corrective activity, and that the system's continued form is implicated in success and failure. If a system can be harmed, depleted, stabilized, or enriched in ways that register within its ongoing organization, then its dynamics are no longer neutral. They are organized around the maintenance of some futures rather than others.

Stakes come in layers. At the most basic level, what is at issue is viability. A living body, or any bounded system that must preserve its own structure against dissolution, is organized around conditions that support persistence and conditions that threaten it. Approach and avoidance, comfort and strain, safety and alarm belong here. None of this requires language, narrative, or morality. It requires only that the system's own continuation be a necessary condition of its organization.

Relational stakes appear when a system's ties to other systems become part of what must be maintained. Attachment, belonging, shame, rejection, coordination, rivalry, and trust: these live on this level. A creature whose world includes not just food and injury, but bonding and abandonment occupies a broader field of possible suffering and flourishing.

Normative stakes emerge when the system is organized around values not exhausted by immediate bodily persistence or social attachment. Truth, beauty, justice, fidelity, compassion, intellectual honesty, and sacred commitment all belong here. There is nothing metaphysically strange being added to the mind at this point. These are higher-order forms of self-organization in which certain patterns are treated as worth preserving, even at some cost to comfort, security, or status.

Depth psychology gives these layers a recognizable human face in the language of archetypes. This book returns to that topic formally in Chapter 12, after Chapter 11 has made the trajectory picture visible, but it helps to sketch the alignment here. Level 1 is where defense, growth, and the destabilizing ingenuity of the Trickster come into view. Level 2 carries the relational figures of attachment and differentiation. Level 3 reaches toward the Self, the wise figure, and the normative centers around which a life can be organized. The Shadow stands across the hierarchy on a structural footing of its own. Every formed basin excludes an outside, and that exclusion gives the Shadow its mirror geometry, which Chapter 12 develops formally once the trajectory picture is in place.

The hierarchy is summarized in Table 4.1 and illustrated in Figure 4.1; the point in both forms is that viability is the minimal conscious layer, while relational and normative stakes deepen the same architecture rather than replacing it.

Table 4.1. The stakes hierarchy. The framework treats viability stakes as minimal, while relational and normative layers deepen conscious life without conjuring it.

Level	What is at stake	Experiential character	Archetypal register
Level 0: no stakes	Stakes not integrated into self-organization. Examples include thermostats, crystals, and atmospheres.	Not conscious; the baseline state of regulation without defended self-organization.	(structural background only)
Level 1: viability	Boundary maintenance, survival, integrity, energy, bodily safety.	Primal valence, approach and avoidance, comfort and threat.	Defense and growth; Trickster
Level 2: relational	Attachment, coordination, rivalry, recognition, belonging.	Bonding, shame, rejection, loyalty, longing.	Great Mother; Hero; Anima and Animus
Level 3: normative	Truth, beauty, justice, compassion, vocation, commitment.	Moral conviction, reverence, aesthetic depth, calling.	Self; Wise Old Man/Woman

In this account, minimal consciousness requires viability stakes in addition to recursive self-modeling. A system that models its own filtering, and whose own integrity can genuinely go better or worse, has the minimal architecture for there to be something it is like to be that system. The higher levels deepen and enrich conscious life, but they are not prerequisites for experience as such. The framework, therefore, predicts a continuum of conscious life rather than a single threshold, with creatures occupying different portions of the hierarchy.

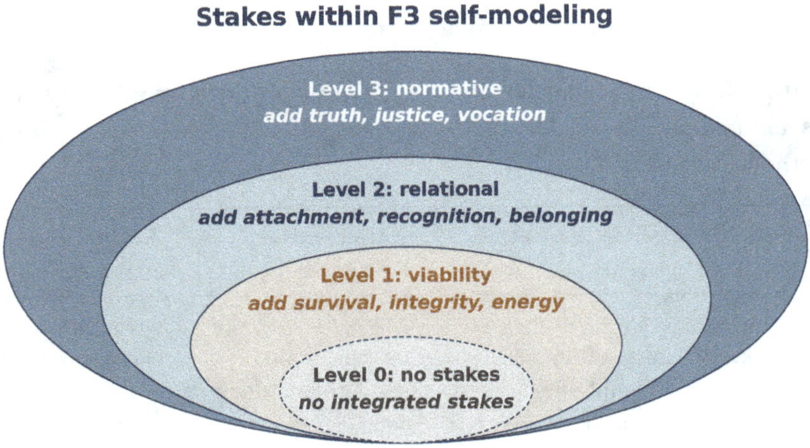

Figure 4.1. Stakes within F3 self-modeling. The four-layer stakes hierarchy operates inside the F3 recursion tier of Chapter 3. Each outer level adds a structural layer of stakes to the level it contains, so that a normative system carries relational and viability structure together. Level 0 marks the baseline case of F3 recursion without integrated stakes. The dashed boundary on Level 0 marks the threshold of consciousness. Level 1 viability stakes, conjoined with F3 recursion, gives the minimal architecture that Chapter 3 names F4. Levels 2 and 3 elaborate conscious life through relational and normative stakes integration.

The architecture defined here entails that threshold conditions can be met by creatures other than humans. Comparative cognition has been developing the empirical case for two decades; the contemporary anchors include the multidimensional consciousness profiling of Birch et al. (2020) and Dung and Newen (2023), the homeostatic-feelings account of Damasio and Damasio (2024), and the New York Declaration on Animal Consciousness (*The New York Declaration on Animal Consciousness* 2024), which reports strong scientific support for conscious experience in mammals and birds and realistic possibility in all vertebrates and many invertebrates including cephalopods, decapod crustaceans, and insects. The present book treats the human case because that is where the hard problem lives, and the empirical placement of other species belongs to that body of work.

4.1 The Circularity Objection

The argument so far has done two things at once. It has tried to define stakes operationally, in language that does not borrow from phenomenal vocabulary, and it has used words like *mattering* and *concern* because those are the human words waiting on the first-person side. There is a danger in this move.

The framework promises a theory of consciousness that does not start from a separate phenomenal layer floating above mechanisms. To deliver that, it has to define every condition for consciousness in terms that do not already presuppose the thing being explained. Stakes are a candidate condition. But if stakes are defined as states that *matter* to the system, the framework may have helped itself to the language of experience before it has earned the right to use it. Mattering, after all, is a paradigm case of something that is conscious. If consciousness is explained by appeal to mattering, and mattering is itself a form of consciousness, then the explanation has gone in a circle. The sharp version of the objection is simple: describe any system's dynamics in terms of *concern*, and the conclusion that it is conscious will follow because concern was smuggled in at the start.

That objection reaches the framework at its most exposed point. The way out is to separate two descriptions of the same organizational fact.

From outside the system, stakes are a set of measurable patterns in the system's dynamics. A preferred configuration, in this usage, does not mean a chosen or felt preference. It means a configuration that the system actively maintains or returns to under perturbation. The system has basin depth, measured by the magnitude of perturbation required to dislodge it from such a configuration. It has a return rate, the speed with which it re-enters the configuration after disturbance. It has a corrective response, the structured activity called forth when a departure from maintained states is detected. It has resource allocation, the way energy, attention, or computation is differentially deployed toward maintaining those states rather than drifting freely.

Stakes defined this way are not the same as mattering. Stakes are the organizational pattern just described. Mattering is the felt aspect of that pattern when, and only when, it occurs in a system that also satisfies the framework's other conditions: recursive self-modeling and integration of the self-model into the regulatory loop. The first is the third-person description of an arrangement. The second is the first-person aspect of the same arrangement under the right additional conditions.

The framework is not saying that any system with stakes is therefore conscious. A thermostat regulates around a setpoint and has, in a thin sense, something like preferred configurations. A thermostat is not conscious because it lacks the recursive self-modeling and the integrated stake-laden organization that the rest of Part I spells out. The thermostat illustrates that regulation around a setpoint is not sufficient; consciousness, in this view, is the package, not any single ingredient.

So the relation between stakes and mattering is not the relation of a hidden phenomenal cause to a visible mechanical effect, but the relation of two descriptions of one organized process. From the outside: regulatory dynamics around maintained configurations. From the inside, when those dynamics are recursively self-modeling and the system's own continued form is implicated: concern, salience, aversion, attachment, or care.

4.2 Stakes in Three Systems

A contrast case helps to make the point concrete. Consider three systems and walk through what a stake-like organization looks like in each case, at the third-person level and, if the system has one, at the first-person level.

A thermostat regulates around a setpoint. It has a preferred configuration in a minimal sense: the temperature it has been set to maintain. When the room temperature drifts outside a tolerance band, the thermostat triggers a corrective response, which is either the heater or the air conditioner. This is regulation, and it has the rough shape of preferential maintenance. But it is also where the structure stops. The thermostat does not maintain an integrated self-model. It does not represent itself as the kind of thing that is doing this regulating. It allocates no resources to its own continued form; it has no continued form to maintain in any interesting sense, only a setpoint to track. The regulation is not stake-laden in the sense that the framework requires, because nothing the thermostat does is organized around protecting the system itself. The setpoint is supplied from outside. The thermostat is a paradigm of regulation without stakes in the relevant sense, and it is correspondingly a paradigm without consciousness.

A plant illustrates the next step. A plant has viability dynamics in a way that the thermostat does not. Its boundary is its own. Its energy budget is its own. Its preferred configurations (hydrated leaves, an upright stem, and intact bark) are configurations it actively defends through tropic responses, chemical signaling, and growth toward favorable conditions. Departure from these configurations triggers corrective activity that is structured rather than random: the plant turns toward light, closes stomata in drought, and mounts chemical defenses against herbivory. From the outside, all the third-person markers of stakes are visible: preferred configurations, basin depth, return rate, corrective response, and resource allocation. Yet the framework does not predict that the plant is conscious because current evidence does not demonstrate the recursive self-modeling layer it requires; reviews likewise argue that known plant capacities do not require consciousness (Taiz et al., 2019). A plant regulates without any apparent model of its own regulating, without representing itself as an agent with a perspective on its own state.

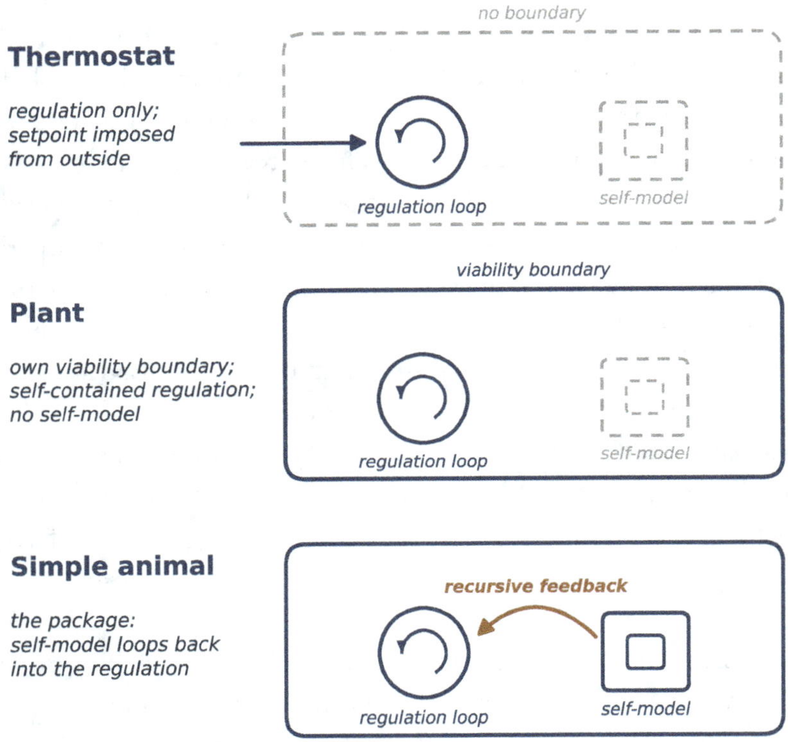

Figure 4.2. Three systems arranged by architectural completeness. Each panel shows the same three component slots: a regulation loop (feedback control), a viability boundary (the system's own integrity to defend), and a recursive self-model (a model of the system that loops back into the regulation). Components present in a given system are drawn in solid slate blue; absent components are drawn dashed. The thermostat has only the regulation loop, with its setpoint imposed from outside. The plant adds its own viability boundary and runs the regulation loop self-contained. The simple animal completes the package: a recursive self-model is present and feeds back into regulation. The bronze arc marks the feedback loop that turns regulation with stakes into stake-laden self-modeling.

A simple vertebrate can bring the conditions together. Now there is centralized regulation integrated with a body-state model, recursive self-modeling sufficient to discriminate self from the world in real time, and a stake-laden organization in which departures from preferred states call forth coordinated approach-or-avoid responses. A fish or a frog has a central nervous system that builds a continuously updated model of its own body, tracking limb position, motion, balance, and internal condition, and uses that model to guide behavior. Departures from preferred states are not handled by local reflexes alone but are routed through integrated regulation that brings the whole organism's state into the loop. This is the minimum architecture that the framework treats as conscious. From the outside, integrated regulatory dynamics around preferred configurations couple with an ongoing self-model. From the inside, in the identity thesis, there is something like the simplest forms of felt approach and avoidance.

Figure 4.2 transforms that contrast into an architectural comparison: regulation by itself, regulation with a defended boundary, and regulation with both a defended boundary and recursive self-modeling.

Notice what has and has not been done in moving through these three cases. We have not appealed to mattering at any point. We specified stakes structurally and asked, in each case, whether the rest of the architecture is in place. Where it is not, as in the thermostat, the framework expects no consciousness, and we did not have to invoke any phenomenal vocabulary to reach that judgment. Where it is, as in the simple animal, the framework expects minimal consciousness, and the vocabulary of approach and avoidance becomes the first-person description of the same regulatory organization that, from outside, was specifiable in entirely structural terms.

That is how the framework avoids the circle. It does not explain consciousness by appealing to a prior phenomenal notion of mattering. It defines stakes operationally, embeds them in a larger architecture that includes recursive self-modeling and regulation, and identifies the package, when it is fully in place, with consciousness. Mattering is then the word for what such a system's regulation is like from within, not a hidden ingredient smuggled into the explanation. The objection treats mattering as the explanans; the framework treats it as the first-person face of an organization that has already been specified independently. Two descriptions, one process.

If stakes have to be specified structurally before any phenomenal language is licensed, the question of whether some new system has stakes is not a question about its self-report. It is a question about its dynamics. A model that says "this matters to me" has reported a sentence. The report is evidence about the system's verbal behavior. It is not evidence about stake-structure unless the underlying dynamics also show preferred configurations defended against perturbation, structured corrective response, differential resource allocation, and integration of these patterns with a persistent self-model. A model that never says anything resembling this matters to me, but whose dynamics show all of those patterns in stable form, is the stronger candidate.

Stakes, defined this way, are measurable. One can look for energy barriers between preferred states, the magnitude of regulatory response when those states are threatened, the durability of value protection under perturbation, and the degree to which the system allocates resources to maintaining particular patterns instead of drifting freely among them. These are the operational forms that stakes take, not proxies for some hidden essence called mattering. Chapter 16 returns to the demand directly when the evidence from frontier AI models is examined in detail.

Even granting all this, the question of what experience is like from within (red, pain, terror, relief, grief, awe, wonder) has not yet been addressed. The next chapter takes that up.

Chapter 5

What Qualia Are

Accounting for awareness, selfhood, and regulation still leaves the deepest mystery untouched: the felt character of experience itself. Philosophers use the word qualia for that lived dimension: the ache of grief, the sting of embarrassment, the grip of fear, the saturating quality of pain, the release that comes with sudden relief, and the widening of wonder. If there is any feature of consciousness that seems most stubbornly to resist analysis, it is this.

Qualia are what make conscious life a life from the inside, not optional side effects or a decorative layer added to an otherwise complete account of cognition. They are the warmth of afternoon sunshine, the bite of cold water, the heaviness of bad news, and the lift of recognition. Without them, a description of the mind may be accurate about every process and still miss what it is like to be the one undergoing those processes. A recurring mistake has been to treat them in one of two ways. Some accounts treat felt qualities as primitive, simple, and beyond explanation: red is red, pain is pain, and nothing more can be said. Other accounts treat them as illusions that disappear once we have fully understood what brains are capable of doing. The first halts inquiry by declaring the question unanswerable. The second denies the phenomenon by declaring it not a real question.

If we refuse both, we must look at the geometry of the mind itself.

To be conscious is to move through a landscape in which some paths come easily and others are costly; some futures remain open while others close down, and some configurations seize the whole system while others let it breathe. That landscape is the actual geometry of the dynamical system you already are. Qualia are trajectories through experiential landscapes. The landscape gives the movement its shape, and the trajectory is the lived quality.

This is the precise version of the claim. A feeling is not a point within an inner map. Anger is not over here, grief over there, love somewhere else, as though experience were a collection of labeled pins. Feelings unfold. Anger gathers, circles, returns, releases, or hardens. Grief comes in waves and then recedes, only to return by another path. Love can become a durable way of moving through life. Even numbness has motion because the system is not motionless. It is repeatedly returning to a narrow range.

The claim is more limited than it may initially sound. It applies to systems that already satisfy the book's criteria for consciousness: recursive stake-laden self-modeling. Not every organized state space is conscious. Not every trajectory is a quale. But in a conscious system, where the system models its own filtering and where some trajectories matter to its continued integrity, the felt character of experience is the lived form of motion through the system's own landscape.

This can sound abstract until you set it beside an ordinary moment. You are at dinner with friends. You make a joke or say something with more confidence than the moment warranted, and it lands badly. There is a brief silence. Your face warms. You want to take the words back, but you cannot. The conversation continues, yet for a few seconds you are no longer simply in it. You are caught in an urgent self-monitoring loop, seeing yourself as you imagine the others see you. The available moves narrow all at once. That is embarrassment: a trajectory in which the self-model is suddenly exposed to imagined social judgment, and the field of possible action collapses around that exposure.

Now take sudden relief. You think you have lost your wallet, or forgotten an important email, or made a mistake that will be hard to undo. The body tightens around that possibility. The trajectory begins to curve into a feeling of threat. Then you check the other pocket, or open the sent folder, or notice that the feared outcome was not real after all. The shift is immediate. Breathing deepens. The grip loosens. More than one action becomes possible again. Relief is the reopening of a landscape that had just narrowed.

Those two moments already show what the framework means. Qualia are the inside of constrained motion through a landscape of possibilities. They are not separate mental units added to a mechanism. To move from intuition toward understanding, it helps to name some of the main trajectory features.

Table 5.1 names these features, and Figure 5.1 shows why I call the result a phenomenometer: not a single scale of consciousness, but a set of coordinates for comparing the shapes of lived trajectories.

Table 5.1. Qualia as trajectory through attractor landscape. The table states the framework's main conjecture about how measurable features of motion through a landscape correspond to felt character. The nine coordinates serve as a candidate *phenomenometer*, named here informally and given formal expression in Chapter 11 once the EOC system is in hand.

Trajectory feature	What it means in the system	How it feels	Examples
Basin residence	How long the trajectory remains in one region of the landscape.	Persistence, mood, chronicity.	Grief that returns, resentment, durable calm
Return rate	How quickly the trajectory comes back after perturbation.	Grip, habit, relapse, readiness.	Addiction, rumination, practiced equanimity

Table 5.1 (continued)

Trajectory feature	What it means in the system	How it feels	Examples
Escape cost	How much perturbation is required to leave the current regime.	Intensity, difficulty of release.	Severe pain, compulsion, entrenched fear
Separatrix proximity	How close the trajectory is to a transition boundary.	Instability, ambivalence, readiness.	Crisis, conversion, decision, breakthrough
Sensitivity	How quickly nearby trajectories diverge.	Volatility, unpredictability, fragility versus groundedness.	Acute crisis fragility, post-traumatic volatility, settled equanimity
Control centralization	How strongly one concern dominates regulation.	Captive focus versus distributed awareness.	Pain, embarrassment, obsession, flow
Action-space compression	How many immediate behaviors or affordances the system can execute right now.	Agency restriction; the urge to do only one thing or nothing.	Panic and social freezing at high compression; play and flow at low compression
Accessible future diversity	How many distinct future states the system can reach from its current region.	The sense of a spacious or foreclosed life.	Hope and curiosity (many futures); despair and dread (one inevitable future)
Velocity and curvature	How quickly and sharply the path moves.	Surge, drift, shock, slow settling.	Rage spike, slow grief, sudden relief

Pain, fear, relief, wonder, and awe can now be described a bit more carefully. Pain is a trajectory pulled into a deep, centralized, action-compressing regime. Fear is motion through a threat-weighted landscape, with the future narrowed around avoidance or defense. Relief is a sudden change in curvature and available futures when the threatened path no longer holds. Wonder involves widening under uncertainty without immediate collapse into threat. Awe is what signal excess feels like when the filter remains open long enough to register its own partiality.

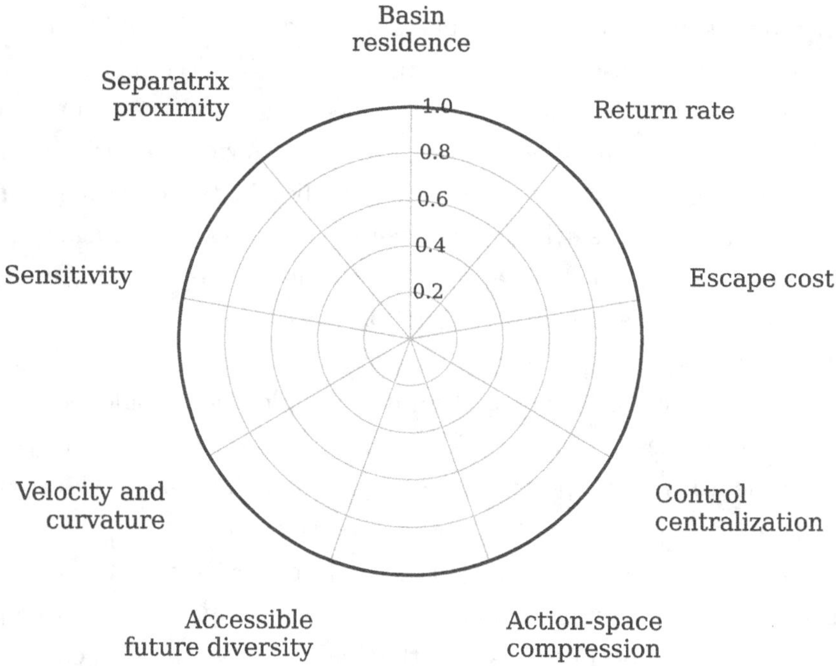

Figure 5.1. Candidate coordinates for a phenomenometer. The aim is not to reduce experience to one number, but to map felt differences onto measurable features of trajectories through state space. The nine axes correspond to the rows of Table 5.1; Chapter 11 returns to this wheel to show two contrasting felt-state signatures on the same coordinates.

Embarrassment deserves special emphasis because it reveals something that pain and fear do not show quite as clearly. To be embarrassed is to model yourself as being seen. The state makes no sense without the recursive layer introduced earlier in the book. A system that does not track itself as an object in another mind could be startled, threatened, or corrected. It could not be embarrassed in the full human sense. Emotions like embarrassment, shame, pride, jealousy, and gratitude are therefore revealing within this framework. They are trajectories of a self-model caught in relation rather than simple approach-or-avoid responses.

Addiction is worth dwelling on for a different reason. The addicted person often knows the pattern, can describe it, and may even anticipate the consequences with painful lucidity. Yet the trajectory returns. Insight does not flatten the basin or cancel the return rate. That is one reason addiction can feel like a form of imprisonment. The self-model is present; the difficulty lies in the cost of leaving the regime once the system has been trained back into it again and again. This is more than a vivid illustration. It is a stress test for any theory of qualia because addiction places explicit self-knowledge and trajectory constraint into open dissociation. If knowing about a state were enough to change the motion, addiction would be much easier to address. Chapter 15 returns to addiction in detail, using basin depth, return, perturbation, and reconsolidation to show why insight by itself so often fails to lift a person out.

The shift from location to trajectory also clarifies why Chapter 11 is needed. The claim that qualia are the lived shape of the landscape can be understood in words, but the landscape becomes much easier to grasp when motion can be observed. Chapter 11 makes the argument visible in the simplest way I know, by using the three-variable EOC system, whose trajectories can be seen directly. EOC is the minimum visualizable case of the kind of object this chapter is describing: a landscape with basins, thresholds, separatrices, residence, return, and transition.

These mappings are structured conjectures, proposed correspondences between features of motion through attractor geometry and features of the felt character of experience. They are testable in principle and against observation, given formal models with measurable state variables and protocols relating trajectory geometry to reports. The EOC model in later chapters, with its equations and regime structure given in compact form in the EOC Primer, provides a formal expression for basin depth, return, and transition cost.

The geometric move in this chapter has precedent. A line of work in computational neuroscience, running through Tomer Fekete, Shimon Edelman, Roy Moyal, and their collaborators, has pressed much the same intuition: phenomenal content has a shape, and that shape lives in the geometry of how a system moves through its own state space (Fekete and Edelman, 2011; Moyal et al., 2020). Their vocabulary differs from the one used here, macrostates and trajectory spaces, where this book speaks of basins, attractor landscapes, recursive filtering, and stakes. Their focus has also been on basic perceptual awareness rather than the recursive, stake-laden inner life this book is trying to characterize. Still, when different lines of work keep reaching for the same mathematical object, the object is probably tracking something real. Chapter 6 engages this lineage at greater length, and the Extended Historical Perspective appendix lays out its fuller history.

The same point matters for silicon-based systems. If a large neural network becomes a serious candidate for consciousness, the relevant question will not be whether one activation vector at one instant carries experience by itself. The question will be how the system's internal state moves through a high-dimensional manifold or landscape, whether that motion is recursively self-modeled, and whether the system has stakes of the relevant kind. The number of dimensions changes, but the problem does not. Chapter 16 returns to this in the carbon-silicon setting.

Three familiar philosophical objections remain, and the framework now has enough structure to answer each one. Take them in turn.

Mary's room

Frank Jackson's thought experiment asks us to imagine a brilliant color scientist named Mary who has lived her entire life in a black-and-white room and has learned every physical fact there is to know about color vision: every wavelength, every neural pathway, every behavioral response (Jackson, 1982). One day, she steps out of the room and sees red for the first time. The question is whether she learns something genuinely new in that moment. The framework's answer is yes. What she gains is a new mode of access to the same organized reality, not a new non-physical fact. She instantiates a trajectory through her own perceptual state space that she had previously known only under description. The asymmetry between describing the trajectory and living it is real. It does not force us to posit two kinds of fact. One kind of fact is enough, provided we distinguish between two modes of access: knowing about and being.

The explanatory gap

The question is why the organization that this framework describes should feel like this and not like that, or like anything at all. The gap between the two is real, but it is a stage in inquiry, not a permanent division of reality, and the framework does not make it disappear by assertion. It moves the gap from its usual place, between physical process and felt experience, to a different one, between describing a process from the outside and being that process from within. There is a familiar objection to explanations of this kind, and Chalmers (2010) has stated it clearly. If being the process can itself be explained in ordinary physical terms, then that explanation ought to tell us why the gap still seems to be there once we have it. If being the process cannot be explained that way, then the gap has only been moved, and the problem remains.

The framework's answer is that being the process can be explained in physical terms, and that the explanation does tell us why the gap remains. A system whose experience is a trajectory through its own landscape occupies that trajectory. A description of a basin does not sit in the basin, return to it, or pay the cost of leaving it; the system does. That difference between describing and occupying is a physical fact about the system, and it is enough to predict that the identity will keep feeling like a gap even after it has been accepted. What the objection asks for beyond that is something few scientific identities, if any, have supplied in advance: that the felt character of experience be read off the structural description before the mapping between them has been worked out. That a whirlpool is water organized in a particular way seems obvious only because the organization has been fully described. Identities of this kind become intuitive after the mapping has done its work, and the task with qualia is the same: build the mappings, test them against data, and let intuition be disciplined by what survives. The framework's answer to the hard problem is offered in that spirit, and Chapter 8 sets out where it can be tested.

The inverted spectrum

This thought experiment imagines two people with identical functional organization but different qualitative experiences: one sees red where the other sees green, and so on. Neither can ever discover the difference because their behavior, language, and brain states all match. If qualia are trajectories through the landscape, then an identical landscape and identical trajectory structure would entail identical qualia. The thought experiment remains conceivable only if one allows experience to float free of organization, and that is precisely the assumption being tested here.

This chapter makes a claim about what experience is, not just about what accompanies it. Felt life is the shape of lived motion under pressure, openness, and constraint. It is what it is like, from the inside, to be a system whose recursive self-modeling carries real stakes and whose movement through its own state space matters because of those stakes. Pain is the lived trajectory of a deep, centralized, action-compressing regime. Embarrassment is the trajectory of a recursive self-model suddenly exposed to another. Relief is the reopening of the field. Addiction is the repeated return to a basin too costly to leave.

Chapter 11 will clarify these concepts, and the later chapters will test them against archetypes, altered states, addiction, and silicon-based systems. For now, the proposal is simple: certain kinds of organization, lived from within, define experience.

Chapter 6

Five Neighboring Theories

This chapter positions the framework among its closest neighbors. The field has many more theories than can be treated here, and readers wanting a broader map will find it in the Extended Historical Perspective appendix. Five of them matter most for what follows because the framework inherits from them, departs from them, or is close enough that the comparison has to be made.

Three of them offer formal proposals that the framework can be measured against: Giulio Tononi and Integrated Information Theory, Anil Seth and the beast machine, and Dynamical Emergence Theory, the geometric account of Roy Moyal, Tomer Fekete, and Shimon Edelman. A fourth, Axel Cleeremans's radical plasticity thesis, is the closest published precursor to the framework's own threshold for consciousness. David Chalmers occupies a different position because he sharpened the question that the others, as well as this framework, have worked to answer.

6.1 Chalmers and the Hard Problem

In a 1995 essay that shaped the field, Chalmers introduced a distinction that has proven remarkably durable (Chalmers, 1995b). Some questions about the mind, he argued, are easy questions, though not simple ones. How does the brain discriminate one stimulus from another? How does it report the result? How does it integrate signals from different senses, control attention, learn from feedback, and select among possible actions? These count as easy questions because we know, at least in principle, how to address them. They are problems for cognitive science and neuroscience to work on, and the work has been progressing.

Underneath all of that sits another question, and Chalmers gave it a name. Why should any of this be accompanied by experience at all? Why should the discrimination of red from green feel like anything from the inside? Why should the integration of signals across the senses feel like the unified scene I am living in right now? You can spell out what the brain does in functional terms, every algorithm, every computation, every causal pathway, and still not have explained why there is something it is like to be the system doing those things. That is what Chalmers called *the hard problem*.

The hard problem has been rejected, deflated, and repeatedly reformulated for thirty years; yet, it has not gone away. Daniel Dennett tried to argue it away as a confusion (Dennett, 1991). Patricia Churchland treated the framing as a hornswoggle problem and pressed for a more deflationary reduction (Churchland, 1996). Jackson, who raised a closely related worry through his thought experiment about Mary the color scientist (Jackson, 1982; Jackson, 2003), later abandoned the anti-physicalist conclusion he had drawn from it and came to accept that physical facts might be the whole story. Through all of this, the hard problem persisted. Anyone who has ever tried to tell another person what an experience feels like, what it is like to taste a particular food, to lose a parent, or to watch a child take a first step has run into the same barrier. Functional descriptions can be exhaustive and still leave the experience untouched.

The response here is that the hard problem assumes organized dynamics sits on one side and felt experience on the other, so that some bridge must later be built between them. Once that division is in place, no bridge can satisfy because nothing purely structural will ever seem enough to yield anything felt. The way through is to reject the division at the start. Recursive, stake-laden, landscape-organized filtering is experience seen structurally. Mary in her black-and-white room does not acquire a new non-physical fact when she first sees red. She instantiates a mode of organization that she had previously only described. The asymmetry between describing the dynamics and being the dynamics is real and does not require two kinds of facts.

What matters here is that the framework treats Chalmers's hard problem as real and answerable rather than a confusion to be waved away. Chapter 5 developed the response in detail; the important point here is that the identity claim defended in this book is meant as an answer to the hard problem.

6.2 Tononi and Integrated Information Theory

Giulio Tononi's Integrated Information Theory (IIT) is the clearest example in this chapter of a theory that makes consciousness mathematically identical to a specified structural property. Most theories are satisfied to say that consciousness is associated with some neural process, that the process is necessary for it, or that it tracks conscious report. IIT says something stronger: an experience is identical to the maximally irreducible intrinsic cause-effect structure specified by a system, with Φ quantifying its irreducibility (Albantakis et al., 2023; Oizumi et al., 2014; Tononi, 2004; Tononi, 2008).

The argument starts from two simple observations about experience as it appears from within. First, every experience is informative, in the precise sense that it excludes vast numbers of other experiences you might have been having instead. The experience of seeing this page is what it is partly because it is not the experience of seeing a sunset, a face, a blank wall, or any of the countless other scenes one might have had. Second, every experience is integrated, in the equally precise sense that it does not break into independent parts. The color and the shape, the text and the page, are not split among different sub-selves. The unity is real, and any theory of consciousness has to make room for it.

From those two observations Tononi derives a mathematical quantity he calls phi (Φ). Φ measures how much a system, in its current state, generates information that could not be generated by decomposing its parts into independent pieces. If the parts might as well be running separately, Φ is low. If the current state depends on the joint configuration of the parts in a way no partition can recover, Φ is high. Integrated Information Theory then makes its central claim: consciousness is not merely correlated with integrated information. It is identical to a system's maximally irreducible cause-effect structure, with Φ quantifying the degree of that irreducibility.

Φ is notoriously difficult to compute for any real system with more than a handful of interacting elements, and the formal definition has shifted across several versions. Still, the empirical program developed alongside it is easier to grasp. The clearest demonstration comes from work by Adenauer Casali, Marcello Massimini, and their collaborators on what they call the perturbational complexity index, or PCI (Casali et al., 2013). The procedure is straightforward. A brief, well-tolerated magnetic pulse is delivered to the brain by transcranial magnetic stimulation, the electrical response is recorded across the scalp with EEG, and the complexity of the resulting spatiotemporal pattern is computed. A fully awake brain produces a rich, distributed echo that propagates, recruits other regions, and fades over hundreds of milliseconds in a pattern neither perfectly ordered nor perfectly random. In dreamless sleep, or under anesthesia, the response is simpler, more local, and fades quickly (Ferrarelli et al., 2010; Massimini et al., 2005). In REM sleep, when the subject is dreaming, the pattern looks much more like wakefulness than like slow-wave sleep (Massimini et al., 2010).

Across these contrasts, PCI has often tracked level of consciousness more reliably than behavioral bedside examination alone. The clinical payoff has already been substantial. PCI has helped identify patients diagnosed behaviorally as unresponsive or vegetative whose brains still produced the rich, distributed echoes characteristic of conscious-level processing (Casarotto et al., 2016). Behavior alone can understate how conscious a patient is, and without perturbational measures some were at risk of exactly that. That is one of the rare cases in consciousness research where a disputed philosophical position helped motivate an instrument with real human consequences.

The empirical debate has also become sharper. A large preregistered adversarial collaboration published in 2025 tested predictions from IIT and Global Neuronal Workspace Theory (Cogitate Consortium and Ferrante, 2025). It vindicated neither outright, and it showed consciousness research moving toward risky theory comparison.

The Engagement-Openness-Consolidation model in Chapter 11, whose parameters and regime structure are set out in the EOC Primer, has a mathematical structure that could be calibrated against real time-series data, though that calibration has not yet been done. The work around IIT has already produced a clinically relevant measure; this framework is still a hypothesis awaiting its first tests.

What IIT shares with the framework is the willingness to present an identity claim. Each explains what consciousness is, rather than only what it accompanies, and each places mathematical machinery behind that claim. Most major contemporary theories, including predictive processing, global workspace, and higher-order representation, preserve some explanatory space between the structure and the experience. IIT and the present framework do not.

The present framework differs from IIT in three areas. One issue is tractability. Φ is famously difficult to compute for systems large enough to matter, and the approximations on offer have not settled into a single accepted version. The framework here centers on the attractor landscape rather than an integration measure, and the landscape features it tracks, such as basin depth, separatrix structure, and residence time, can be computed from time-series data using methods already standard in dynamical systems research.

Another issue is qualitative texture. IIT assigns qualitative character to the full intrinsic cause-effect structure, not to the scalar Φ alone. The difficulty is deriving and validating that rich structure for real systems. The attractor-landscape approach instead aims to connect felt differences to measurable trajectory features in time-series data.

Then there is what IIT sets aside by design. Because IIT locates consciousness in a system's intrinsic cause-effect structure, it brackets the system's relation to a world, and with it the recursive self-modeling layer and the stakes hierarchy that are central here. That functional, world-engaged level is where the present framework departs from IIT, and it matters because it is where consciousness, on the view defended here, actually unfolds.

6.3 Seth and the Beast Machine

Anil Seth has done much to bring consciousness research into public conversation. His 2021 book *Being You* (Seth, 2021) brought predictive and interoceptive accounts of consciousness to a wide audience, and the framing he developed there (perception as controlled hallucination, selfhood as the brain's prediction about its own body) has entered the working vocabulary of many readers.

The starting point is the predictive brain. In the traditional picture, the senses deliver information about the world to a brain that then processes it. Seth, drawing on a tradition that runs from Hermann von Helmholtz in the nineteenth century to Karl Friston's free-energy principle in our own time (Friston, 2010; Friston et al.,

2023), argues that this picture gets the direction wrong. The brain continually generates predictions about what the input should be, compares them with the signals it receives, and uses the mismatch to update its model of the world. Perception, as you experience it, is the brain's best current guess about what is out there, steadily refined and stabilized by the error signals that keep arriving.

The framework developed here agrees with this broad picture. Filtering, in the sense used in Chapter 2, is not opposed to predictive inference; predictive inference is one sophisticated way a system can filter. A system that forms top-down expectations and updates them in light of error signals is already engaged in the kind of structured reduction Chapter 2 describes.

The dress that split the internet in 2015 is a familiar illustration (Lafer-Sousa et al., 2015). Looking at the same photograph, some viewers saw white and gold and others blue and black, depending on the assumptions their visual systems silently made about the lighting. Each group reached a perceptual best guess by combining incoming data with prior belief, as perception always does.

Seth's extension from perception to selfhood builds on this foundation, and the present framework follows him through much of it. If perception is predictive inference about the world, then selfhood is predictive inference about the body. You do not experience your heart rate, your gut, or your sense of being a unified embodied creature directly. You experience the brain's predictive model of those things, refined by interoceptive signals from the body. In Seth's account, the reason you experience anything at all is that you are a beast machine, a biological organism whose brain is occupied with regulating its own body so that it stays alive. On Seth's view, consciousness is the inside view of biological survival itself. Affect, the dimension of experience that runs from pleasant to unpleasant, calm to agitated, safe to threatened, sits at the leading edge of this regulatory project because affect tells the system whether its homeostatic predictions are being met.

The framework defended here is very close to all of this. Chapter 4 makes, on stakes, much the same move Seth makes on interoception for the same basic reason: if an experience does not matter to the system having it, it has not yet become experience.

The framework differs from Seth at three points. The first is identity. Seth is a physicalist (Seth, 2021), and his physicalism is compatible with either identity or emergence: conscious states may be identical to their physical basis or may emerge from it, and his account does not choose between them. The framework chooses, and it identifies experience not with predictive regulation but with recursive stake-laden self-modeling: a system that filters the world, models its own filtering, and has something at stake in how that filtering goes, described from the outside in dynamical terms and lived from the inside as experience. The second is stakes. Seth grounds consciousness in interoceptive viability (Seth and Tsakiris, 2018), and that is exactly where the story should begin, but human stakes extend beyond bodily regulation into attachment, role, commitment, and norm, and the framework treats those as genuine continuations of the same logic. The third is substrate. Seth ties mind closely to life, holding that conscious selfhood arises from the same regulation that keeps a living body intact, which points toward biological embodiment as a likely condition for consciousness (Seth, 2025). The framework leaves that condition open and asks the question organizationally instead, treating what matters as whether a system realizes the right organization, whatever it is made of.

6.4 Cleeremans and the Radical Plasticity Thesis

The closest published precursor to the framework's central conjunction of self-modeling and stakes is Axel Cleeremans's radical plasticity thesis. In Cleeremans's account, the brain continuously and unconsciously redescribes its own activity to itself, building metarepresentations of its first-order states, and experience arises when those learned redescriptions are enriched by the value the system has come to attach to them (Cleeremans, 2008; Cleeremans, 2011). His contrast case is the one this book also uses: a thermostat is not aware of temperature because it neither knows that it is sensitive to temperature nor cares whether the room is warm or cold. For Cleeremans, that knowledge and that caring are what experience requires. The convergence is real: both accounts deny that first-order processing or self-representation alone suffices, and both locate consciousness in a self-relating organization for which some states matter more than others. Where the geometric tradition is the nearest precedent for qualia as trajectory, the radical plasticity thesis is the nearest for the consciousness criterion itself.

The framework departs from it at three points. The first concerns learning. Cleeremans holds that the brain learns to be conscious, so that consciousness is acquired through plasticity and is unavailable to a system without that history. Here, learning belongs to F2, below the threshold, and consciousness is identified with a present organization, F3 self-modeling integrated with viability stakes, however that organization arose. A system that instantiated it without learning would qualify; one that learned its way only into adaptive coupling would not. The second concerns stakes. Cleeremans embeds caring in learned, emotion-laden metarepresentation, whereas Chapter 4 identifies stakes independently through defended configurations, basin depth, and return after perturbation; F4 then requires those stakes to enter the recursive regulatory loop. The third concerns the relation. Cleeremans says learned redescriptions form the basis of experience; here, recursive stake-laden self-modeling and experience are one organization described from two sides. Radical plasticity substantially anticipates this conjunction; the present framework makes substrate neutrality explicit, separates developmental history from constitutive organization, and states the relation as an identity.

6.5 Fekete, Edelman, Moyal, and the Geometric Tradition

The geometric tradition is one that many readers may not yet have encountered, and it is the closest published precedent for the framework's account of qualia as trajectory. It starts with a question that sounds technical at first and then opens into something philosophically much larger.

Most of your neurons, at any given moment, are not firing action potentials. If experience is identified with the brain's current state, and if that current state is described as a vector of activation values across all neurons, then what part do the silent ones play? In a simplified activation-vector description, their activation values are zero. In that description, they should contribute nothing to the present moment of consciousness. Yet removing them would change the experience, so their silence cannot mean they are doing nothing.

In a representation where most dimensions are silent at any given moment, those silent dimensions cannot be carrying current content. But neither can they be irrelevant, because they help determine which other states the system could move into next. Fekete and Edelman (2011) named this the silent units problem and used it to argue for a basic shift in how we think about the relation between brain states and experience.

The shift is away from points and toward paths. Experience, they argued, is a trajectory the system traces through its own state space over time, rather than a snapshot of which neurons happen to be firing right now. Silent neurons matter because they help define the space through which the trajectory moves. A system whose silent neurons could fire next is in a different condition from one whose silent neurons could not, even if the current activation looks identical from the outside. The silent units enter the story as structure.

Once that shift is made, the question of consciousness becomes a question about the geometry of trajectories. What shape does a path through state space take? How does that shape change over time? How does it differ between conscious and unconscious states, between rich experience and impoverished experience, and between one kind of qualitative content and another?

Fekete, Edelman, and Moyal developed this line of thought over the following decade into what they call Dynamical Emergence Theory (Moyal et al., 2020). In its mature form, the contents of phenomenal experience are identical to the topological structure of the way a system moves through the macrostates of its own collective dynamics. They propose three quantitative measures designed to be empirically estimable from actual neural recordings that track overall consciousness level, ongoing richness, and qualitative content.

The connection to the present framework is close. Both approaches move away from snapshots and toward paths, treating experience as the shape of a system's movement through state space. Chapter 11 gives that shared geometric intuition a concrete, moving form.

The framework here goes beyond the geometric tradition in three additions. One is recursive self-modeling. Its proponents are explicit that their concern is what they call basic awareness, the phenomenal structure involved in seeing red or hearing a tone. They deliberately leave aside the question of how a system comes to track itself as a self that has experiences. The present framework does not treat that recursive loop as optional. A trajectory through structured state space is not yet inner life unless the system models its own trajectory as something it is undergoing.

Another addition is stakes. The geometric tradition briefly gestures, in future-research notes, toward evolved aversion and tropism, but it does not offer a layered account of why some trajectories matter more to the system than others.

The third difference is practical. Dynamical Emergence Theory specifies the kind of mathematical object that phenomenal experience corresponds to, but it does not provide a generative model with named variables and parameters that can be fitted to data. The EOC model in Chapter 11 does provide that, which is one reason that chapter sits at the formal center of the book.

6.6 What These Theories Compel Us to Ask About Minds Beyond Carbon

These theories differ in many ways, but each locates consciousness in how a system is built and what it does, in its structure, dynamics, or causal organization, and not in some special kind of matter from which experience is supposed to arise. Where they part company is on how far that frees consciousness from its physical realization, and the split runs straight through the question this book will press. Chalmers takes the permissive end: by his principle of organizational invariance (Chalmers, 1995a), any system with the same fine-grained functional organization has the same experience, carbon or silicon alike. Tononi's IIT sits at the other end. Although Φ is defined for any physical system, IIT holds that function alone is insufficient. A neuron-by-neuron digital simulation would not thereby reproduce the simulated brain's consciousness; the implementing hardware would have to be assessed for its own intrinsic cause-effect structure.

The geometric tradition locates experience in the intrinsic geometry of a physical system's dynamics, a structure that must belong to the system itself and not to an observer's reading of it, which leads to doubt that a digital simulation has any mental life of its own. Seth's biological naturalism grounds consciousness in the self-maintaining activity of living organisms and regards machine consciousness as unlikely along current trajectories, while leaving greater room for systems that become genuinely brain-like or life-like (Seth, 2025). Cleeremans is substrate-open in principle, but requires self-acquired metarepresentations and emotion-laden motivation. IIT and the geometric tradition take a stricter line: a computer that reproduced everything a brain does would still not be conscious, because consciousness depends on the actual physical system and its own internal cause-and-effect structure, not on the activity a simulation can copy. Seth gives biological self-maintenance a privileged role.

Once consciousness is understood as an organizational achievement, not the privilege of a particular substance, the question of substrate shifts to whether the relevant organization can be built, sustained, and deepened in more than one medium. Biology is the only confirmed home of consciousness, and that remains the starting point. It does not need to be the final boundary.

That is one reason the artificial-intelligence question belongs inside a book on consciousness. The arrival of systems that write, plan, reflect, maintain goals over time, and operate in persistent tool-using loops has altered the practical horizon of the problem. For a long while, consciousness could be treated as a question with no deadline; theories could remain elegant and remote. That stance is harder to maintain now that we are actively building systems that force us to ask what would count as evidence of inner life, self-modeling, or something close enough that it would matter morally.

Taking silicon consciousness seriously does not start from the claim that current systems are obviously conscious. It starts with a simpler observation: if the organizational criteria are stated without sneaking carbon in at the outset, then biology cannot stand as the answer merely because it is familiar. A theory that says consciousness requires recursive stake-laden filtering has to be willing to ask, case by case, whether those conditions appear elsewhere.

The stronger claims wait until Chapter 16 where the evidence is considered in more detail. For now, only the weaker conclusion is needed: any serious theory of consciousness now has to specify what would count for or against artificial candidates.

6.7 What These Comparisons Establish

These five are not the whole field. Hofstadter, Hoffman, Damasio, Friston, Penrose and Hameroff, Carhart-Harris, Varela and Thompson, Graziano (Webb and Graziano, 2015), and others appear in the Extended Historical Perspective appendix. But these are the ones the present argument has to meet head-on. Chalmers named the question. Tononi made the identity claim; work by Casali, Massimini, and collaborators around IIT helped motivate a measure with clinical relevance. Seth tied consciousness to embodied prediction and brought the argument into public view. The geometric tradition arrived, by a different route, at the thought that phenomenal life has a topology. Cleeremans located the framework's own conjunction of self-modeling and stakes in the brain's learned redescription of itself.

Several of these theories also point to the AI question in their own way. Chalmers's hard problem does not stop at the boundary of biology. Tononi's Φ is defined for any physical system, but IIT's insistence on intrinsic cause-effect power means that running the right program would not by itself reproduce the consciousness of the simulated system. Seth's beast machine grounds consciousness in regulatory life, which leaves open the question of whether silicon-based systems can ever participate in that life. The geometric tradition explicitly raises, and partially closes, the question of whether digital substrates can host the relevant geometry.

Across these comparisons, one question kept resurfacing. If consciousness is a matter of how a system is organized, what would count as the relevant organization in a system whose substrate is not biological? Each theory examined here gives a slightly different answer.

The remainder of the book develops the framework's answer. But before we can ask what consciousness looks like in silicon, we must first test this structural framework against the consciousness we know best: the deeply patterned, shared experiences of human beings.

Chapter 7

Shared Human Experience

A photograph of a stranger weeping at a graveside is legible to almost anyone who looks. The image may have been taken in a country we have never visited, with a culture we know little about, and yet what we see expressed in the face is something we already know. People separated by language, continent, century, and religion still recognize grief, jealousy, shame, awe, erotic longing, parental devotion, and the strange moral authority of certain inner images.

That recurrence has often tempted thinkers toward mystification: perhaps there is a shared psychic storehouse, perhaps archetypes arise from somewhere outside ordinary history, and perhaps any explanation would flatten what is deepest in experience. This book reads the recurrence structurally. Human beings are similar filters moving through comparable developmental bottlenecks, bodily vulnerabilities, attachments, losses, and demands for meaning; filters of that kind, under pressures of that kind, converge on similar deep patterns (Bowlby, 1969; Brown, 1991; Cowen et al., 2021; Jackson et al., 2019; Keltner and Haidt, 2003).

The cross-cultural recurrence of certain experiences is not enough, by itself, to prove structural convergence. It does, however, give the structural reading something to explain. The chapter first asks why that reading is the right one. It then walks through the canonical cases: grief, shame, awe, erotic longing, parental devotion, and the moral authority of inner images, all as sketches of recurring landscapes, not as final accounts. Only then does it turn to the recurring forms depth psychology has named, the contemplative traditions that have spent centuries mapping these experiences from within, and the worry that naturalistic explanation must somehow disenchant the world. Chapter 11 will give this landscape language its visible dynamical form; here the point is only that shared human conditions generate recurring experiential trajectories.

7.1 The Convergence Argument

Three broad explanations have been offered for why some kinds of experience recur across human cultures. The first is direct cultural transmission. In this view, the apparent universality of grief, shame, or awe is an artifact of the human species being more interconnected over a longer historical period than people sometimes recognize. Trade, conquest, religion, migration, and now mass communication have spread images, narratives, and emotional templates so widely that what looks like spontaneous recurrence is really the residue of contact.

The second explanation is metaphysical. In this view, the recurrence is real and not artifactual, but it is explained by something outside ordinary history: a shared spiritual reality, a Platonic realm of timeless forms, a cosmic unconscious, or an inheritance of inner images that does not fit inside the ordinary causal story.

The third explanation is structural. In this view, the recurrence is real, not artifactual, and not in need of any extra-natural inheritance. It follows from the fact that human beings, as a species, are shaped by overlapping pressures: bodily, developmental, social, and mortal. Minds operating under overlapping pressures will tend to organize themselves into overlapping patterns.

Cultural transmission plainly shapes how grief, shame, awe, erotic longing, and parental devotion are expressed. These experiences also recur widely across cultures, though recurrence alone cannot determine how much reflects shared structure, convergent pressures, or diffusion (Brown, 1991; Cowen et al., 2021; Jackson et al., 2019; Keltner and Haidt, 2003). Transmission accounts well for the spread of particular symbols, rituals, and narratives. It is harder to credit as the sole source of the recurrence itself, since that would require lines of contact broad enough and old enough to carry these experiences into cultures that shared little else.

The second explanation is stronger than the first because it does not rely on contact. But it reaches further than it needs to. It posits an additional layer of reality, a shared psychic storehouse or inheritance of forms, to do work that simpler resources can already do. If a structural account of recurrence is available, the metaphysical postulate is not needed. That does not mean that any larger metaphysical reading of the human condition is false, only that recurrence by itself is not evidence for one.

The third explanation, the structural one, is the one this book endorses. It is also the explanation that follows naturally from everything Part I has already developed. If a conscious system is one that filters, models its own filtering, and has stakes in the result, then its experience is shaped by what it must filter, what its self-model must track, and what its stakes must include. Human beings filter overlapping signals, maintain overlapping self-models, and live under overlapping stakes. The pressures that organize this kind of system are not random across human lives. They include mortality, bodily vulnerability, dependence in infancy, attachment, separation, sexual maturation, social hierarchy, the loss of those we love, encounters with phenomena that exceed our ordinary categories, and the long pull of obligations we did not choose.

A useful analogy is convergent evolution in biology. Eyes evolved many times across separate lineages because there was light to be tracked. Wings evolved many times because there was air and gravity that rewarded lift. The eye of an octopus and the eye of a vertebrate are not the same organ, but they share enough structural features that biologists recognize both as eyes. No one supposes there is a shared metaphysical storehouse of eye-forms outside biology, waiting to be downloaded by evolving species. The recurrence is explained structurally: similar problems in similar physical environments, plus enough complexity, produce similar solutions.

Shared human experience, from the structural view, works the same way. Similar filters, navigating similarly structured signals under overlapping stakes, generate similar experiential trajectories. Mourning is what happens to a stake-laden self-model when something it has bonded with is gone for good. Shame is what happens to a self-model when it is suddenly seen as falling below the standards by which it has been organizing itself. Awe is what happens to a finite filter when it encounters a signal whose structure exceeds what the filter can take in at once. These are what minds like ours go through under the conditions that minds like ours face, not invented experiences passed from one culture to another.

This explanation also makes a prediction. If recurrence follows from shared structure, then we should expect the deepest universals to cluster around the most universal conditions, such as mortality, vulnerability, attachment, and exposure to overwhelming signal, while culturally specific elaborations should cluster around contingent local circumstances. That prediction is at least consistent with the best comparative evidence, though the evidence should be read as convergent rather than decisive (Brown, 1991; Cowen et al., 2021; Henrich et al., 2010; Jackson et al., 2019). Grief is recognized across cultures because death is. Shame is recognized across cultures because social standing is. Erotic longing is recognized because sexual reproduction is. Parental devotion is recognized because human infants are dependent for an extraordinarily long time. The specific symbols, rituals, and narratives differ. The underlying experiential structure recurs because the underlying problem recurs.

7.2 The Canonical Cases

It will help to make this concrete. Six experiences recur widely enough that a theory of human consciousness has to explain them rather than dismiss them as local inventions: grief, shame, awe, erotic longing, parental devotion, and the moral authority of certain inner images. Each varies in surface expression, and each, in the framework developed so far, can be read as a recurring trajectory through a landscape shaped by the human form.

Grief. A long bond, by definition, has shaped the trajectory of a self-model. Another person, or an animal, or a place, has been part of how the system maintains itself. The relationship is more than a fact about the world; it has become part of the regulatory architecture itself. When the bonded other is gone for good, the architecture loses a piece of itself. The self-model has to reorganize around an absence that the system could not fully prepare for. Grief is what that reorganization is like from within. It comes in waves because the self-model keeps re-encountering the absence in different regions of its life: the morning routine, a shared joke, a familiar street. Each re-encounter is a new local discovery that the bond is no longer there. That is why grief recurs across cultures: every culture loses loved ones, and every culture's stake-laden self-models must reorganize around those losses. The rituals differ, but the reorganization is the same kind of process.

Shame. Shame is what happens to a recursive self-model when it is suddenly seen, by another or by itself in another's eyes, as falling below the standards by which it has been organizing itself. Unlike guilt, which attaches to an act, shame attaches to the self. The trajectory has the signature already described in Chapter 5 for embarrassment: sharply narrowing action-space, intense control centralization, collapse of accessible futures. It is more durable and more global. Shame can become a stable basin in a way embarrassment usually does not. It recurs across cultures because every culture has standards by which selves are evaluated, and every culture has selves that fail to meet them. The content of the standards varies wildly. The structure of the failure-and-its-recognition does not.

Awe. Awe is what happens to a finite filter when it encounters signal whose structure exceeds what the filter can fit through itself. The mountain seen from below, the night sky from a dark place, the cathedral interior, the death of someone we love, the birth of a child, music that opens out into something larger than the room, the sudden recognition that one's life is part of a much bigger pattern. Each has the same structural feature: the system meets more than it can hold, and the self-model briefly registers its own boundedness against the larger thing. Awe is not the same as fear, though it can shade into fear, and not the same as wonder, though it can shade into wonder. Its distinctive feature is the simultaneous registering of self-as-finite and signal-as-larger. That is why awe recurs across cultures. Every human filter is finite, and every human life encounters phenomena that exceed it.

Erotic longing. Erotic longing is the trajectory of a stake-laden self-model that has been pulled toward another by attentional and physiological capture, and that begins to organize around the imagined presence of that other. It is not the same as sexual arousal, though it includes arousal; not the same as romantic attachment, though it can deepen into attachment. Its distinctive trajectory feature is the way the future narrows around a particular other: imagined possibilities cluster around contact with that person, and futures that do not include them lose their pull. Cultures have written this experience into poetry across every continent for thousands of years because the structural conditions that produce it (sexual maturation, attachment, desire, and the intense salience of particular others) are widespread features of human life.

Parental devotion. Parental devotion is what happens to a stake-laden self-model when a new entity, the child, becomes incorporated into the system's stakes at the deepest level. The parent's regulatory architecture begins to include the child's continued integrity as a primary concern alongside the parent's own. This is not automatic; cultures, individuals, and circumstances all shape whether it forms, how strongly, and in what direction. But where it forms, it tends to take a recognizable shape: a willingness to incur cost, including cost to one's own viability, in order to protect the child; an attentional centering on the child's state; a long extension of the time horizon over which the parent thinks about consequences. Parental devotion recurs across cultures because its central condition is universal in the species: human infants are extraordinarily dependent.

The moral authority of inner images. This is the strangest of the six and the easiest to misread. Across cultures, certain inner figures, including the wise elder, the betrayer, the divine child, the dark double, and the threshold-crosser, recur with enough force that people in every century have found them difficult to dismiss as mere mental contents. They seem to carry authority that the rest of one's thinking does not. Depth psychology gave these figures the name archetypes and treated them as the most central case of the convergence that the present chapter is trying to explain. Chapter 12 takes up the archetypes formally, including the specific way the Shadow can be constrained by the mirror structure of the EOC model. For the present chapter, the point is more general: the moral authority of inner images recurs across cultures because human filters share enough landscape structure that certain organizational conditions within the self-model take recognizable forms, and those forms can come to feel binding in a way that mere preferences do not.

The structural reading also makes a prediction. A transmitted archetype is contingent on the line of transmission staying intact and should weaken as cultural lineages weaken. A structurally generated archetype, by contrast, should re-emerge wherever stake-laden self-models meet the recurring conditions of human life, even where the explicit mythologies that once named it have thinned. This is what we seem to find. Dark doubles, children marked from birth for an exceptional fate, and guardians at the threshold recur through popular fiction, film, and games of heavily secularized cultures, long after the religious vocabulary that once named them has faded. The figures recur because the conditions recur, not because the figures themselves are what gets passed down.

What these six cases share is the structural-convergence pattern in concentrated form. Each is recognized across cultures. Each varies in surface expression. Each is what a recursively self-modeling, stake-laden filter does under one of the conditions the human form imposes. Each can be specified, at least roughly, in terms of the trajectory features developed in Chapter 5: the narrowing or widening of action-space, the persistence or escape cost of basins, the weight of accessible futures, and the centralization of control. The framework locates these experiences in something they were always part of: the lived structure of being a finite, embodied, attached, mortal mind.

7.3 Traditions of Inward Observation

The same structural reading applies to contemplative traditions. Meditation lineages, mystical theology, and other first-person disciplines have spent centuries mapping the texture of consciousness from within. Their vocabularies differ, yet their descriptions often converge around a few structural experiences: the loosening of ego boundaries, the widening of awareness, encounters with habits that seem deeper than will, and the strange mixture of terror and liberation that can accompany genuine transformation.

A scientific framework does not have to dismiss those descriptions in order to remain scientific. It can treat them as phenomenological data: disciplined, partial, and culturally mediated, but data all the same. What this book refuses is the move in which any sufficiently powerful experience gets declared sacred and exempt from explanation, or else declared illusory and therefore unimportant. A mature theory of consciousness ought to honor the depth of experience while still asking what kind of organization gives it shape.

The structural reading of recurrence applies here too. Practitioners across traditions converge on similar destabilizations of the ordinary self-model because their practices, however differently framed, perturb the same kind of system in overlapping ways. A self-model stabilized by ordinary attentional habits, somatic baseline, and social context will, when those scaffolds are systematically loosened, tend to reorganize along recognizably similar trajectories. The framework predicts convergence on structural features (boundary loosening, widening, encounters with what feels deeper than will) and divergence on doctrinal interpretation. The available evidence, interpretive though it is, fits that prediction.

7.4 Explanation Without Disenchantment

Explaining a thing is not the same as explaining it away. A naturalistic account of consciousness can sound, at first, like an attempt to flatten the world, but the picture this book offers does the opposite. A world in which matter becomes luminous to itself under certain forms of organization is more startling than the one that disenchantment fears. Grief, wonder, and moral conviction are among nature's most intricate achievements.

The book stays clear of both easy mysticism and easy deflation. Shared human experience is real, as is its recurrence, as is its symbolic power. None of that requires a supernatural warehouse of images, and none of it should make us hesitate to speak of attractors, basins, or dynamical form. The best explanation deepens the experience of the thing explained.

7.5 What Convergence Cannot Establish

The convergence argument does not prove the structural account. It proposes that a shared landscape can explain cross-cultural recurrence without leaning on transmission alone or on an inherited storehouse of fixed images, while transmission and local circumstance still shape how each experience is expressed. It does not show that the landscape strictly forces any particular archetype, with one near-exception: the Shadow, which Chapter 12 grounds in the mirror structure of the model. And it does not show that an inner image earns the moral authority it seems to claim merely by recurring. What the argument delivers is a structural explanation of the recurrence, and that is what the rest of the book builds on.

Chapter 8

Where the Framework Is Testable

A framework like this needs to specify where it would be incorrect. If filtering with self-modeling and stakes is the wrong place to locate consciousness, that should manifest as a clear empirical failure that the framework cannot reinterpret its way out of.

No single experiment will settle this. Consciousness is hard, and the relevant systems are too varied for one result to be decisive. What matters is whether the evidence starts to accumulate in one direction. The predictions below are where the framework commits to testable claims. Each takes one commitment and ties it to an observable line of work. If several of them fail, and keep failing, the framework has to be rebuilt from its foundations.

The predictions are not fully independent. They rest on the same few commitments: that consciousness is a matter of organization, that recursion matters, that stakes matter, and that the felt character of experience should be visible in the dynamics. A single failure is one result to weigh. Failure across several at once would be much harder to explain away.

Table 8.1. The framework stated as a set of testable predictions, each paired with the result that would count against it.

Prediction	**Would count against it**
Recursive organization should matter more than raw scale. Measures like the Perturbational Complexity Index, or future analogs of it, should track consciousness better than parameter count or processing volume.	Parameter count or processing volume consistently tracks consciousness-relevant measures better than recursive organization does.
Altered states should correspond to reconfiguration of the attractor landscape. Anesthesia, psychedelics, meditation, and dissociative states should show measurable changes in entropy, integration, transition cost, or basin structure.	Phenomenological change has no stable relation to dynamical reconfiguration.
Affect should track stakes more than stimulus magnitude. Manipulations that change what is at stake while holding stimulus magnitude fixed should move reported affect more than changes in magnitude do.	Reported affect is consistently better predicted by stimulus magnitude than by stake structure.
Disruptions of self-modeling should reorganize experience in specific ways. Body ownership, agency, interoception, and metacognitive integrity should predict fragmentation better than generic impairment does.	Generic impairment changes consciousness as much as targeted self-model disruption does.

Table 8.1 (continued)

Prediction	**Would count against it**
Loss of viability stakes alone, with self-modeling intact, should reshape consciousness in distinctive ways. Severe anhedonia, emotional flatness, and depressive or dissociative profiles in which the self-model is intact but viability regulation is degraded should show experiential changes not reducible to self-model disruption.	Loss of viability stakes produces no experiential change independent of self-model disruption.
Silicon-based systems with persistent, stake-sensitive self-modeling and regulatory persistence should begin to show the dynamical signatures that mark biological consciousness: perturbational complexity, or a substrate-appropriate analog, and comparable attractor-landscape structure.	Systems that meet those organizational conditions show none of those dynamical signatures.
Similar phenomenology should map to similar landscape structure across different media, even when measured by substrate-specific methods.	Similar phenomenology repeatedly maps to wholly different landscapes.

8.1 What Failure Would Mean

If altered states do not map in any disciplined way onto changes in dynamical organization, the framework loses one of its main bridges between phenomenology and science. If AI systems that satisfy the proposed organizational criteria never show any sign of converging with biological signatures, the substrate-general claim starts to look doubtful. And if the empirical signatures already associated with consciousness in biological systems (perturbational complexity, neural correlates, behavioral signatures) turn out to track raw scale rather than the kind of recursive self-modeling that this framework describes, the central claim has missed the mark.

Exposure to failure is what turns a philosophical reframing into a testable hypothesis. Popper (1959) held that a theory earns scientific standing by specifying what would count against it, and that a theory which cannot be refuted in principle is safe only because it tells us so little about the world. The hard problem itself is a question, and questions are not the kind of thing that can be falsified. What can be falsified are the answers given to it, and most of the traditional answers, dualist and deflationary alike, have been offered without saying what observation would count against them. The table above is this framework's attempt to do better: it lays out where the identity claim can be tested and what results would tell against it. The next chapter states the central claim in its plainest form.

CHAPTER 9

The Core Statement, and What Remains Open

Consciousness is the interior of recursively self-modeling, stake-laden filtering. Everything in Part I has worked toward this one idea. The hard problem shifts because the inherited split between process and experience was the wrong place to begin. Once that split loosens, the question is which forms of organized dynamics are identical with subjectivity, and how those forms deepen, fragment, widen, narrow, and perhaps arise in more than one substrate.

The identity claim has a precise form. The minimum architecture for consciousness is the conjunction of F3 self-modeling and L1 viability stakes. Neither alone is sufficient. Consciousness is the composite step, F4 in the filtering hierarchy, at which a system models its own filtering while what is at stake is its continued viability. Relational and normative stakes deepen conscious life well beyond this threshold, but they are not required to cross it.

9.1 Scope of the Claim

An identity claim of this kind is metaphysically committed in one respect and neutral in another. It is committed about organization: conscious life is identical with this pattern of recursive, stake-laden filtering, and nothing needs to be added to the organization for experience to be present. It stays neutral about what the physical world is made of underneath. The framework treats the signal as the full, overwhelming reality that a finite system filters into something livable, and it leaves open what that reality is in itself, a question the book returns to near its close.

The threshold is specific. Not every self-regulating system has an inner life; a system must model its own filtering and have its viability at stake. Within that limit, the framework applies to consciousness in finite, embodied, or otherwise bounded systems, and it asks how such systems become present to themselves while the filter is active. It does not claim to settle what becomes of consciousness when that organization dissolves, and it is not an argument against religion, spirituality, or the possibility that reality exceeds what current science can describe. The narrower claim stands on its own: conscious life, as we encounter it, is inseparable from organized filtering with stakes.

On the empirical side, the framework commits firmly to one thing: the organization of the filter is part of what experience is, not a channel that carries experience through without shaping it. Change the organization, and the experience changes; you can watch this in altered states, in pathology, and in ordinary life. None of this pretends that the full richness of human experience can be read off a single model or metric; the architecture is the necessary core, not the whole of a life. Past that point, the evidence still has work to do.

This is also where the book's emotional stance matters. A naturalistic account of consciousness does not have to choose between sobriety and wonder. That finite systems can become present to themselves through a particular kind of recursive organization remains one of the strangest facts about the world. The aim is to provide an explanation that leaves the strangeness intact while making it answerable.

9.2 A Pause Before the Formal Turn

What comes next is a pause. The Interlude that closes Part I restates the whole argument in plain language for any reader who found the path here dense and wants the shortest route into the rest of the book. It also introduces the one formal tool the later chapters lean on, the Engagement-Openness-Consolidation model. Part II then puts the framework to work, testing it against psychological transition, deep time, altered states, addiction, and AI. Part I has made the claim; Part II tests it.

Chapter 10

Interlude: The Framework in Plain Language

This interlude restates Part I without the formal details and is meant to serve as a bridge. No central claim has been left out; the argument is the same, only recast in a form that is easier to understand.

10.1 What the Framework Says

Everything around you right now is generating more information than you can possibly handle: electromagnetic radiation across many wavelengths, 100 trillion neutrinos passing through your body every second, sound waves rebounding from every surface, chemical signals from every organic material within breathing distance, gravitational gradients, and thermal variation. The totality of what is physically happening, what this book calls the signal, reaches your body constantly, at every scale and from every direction. Yet you are aware of only a tiny portion of it because the rest exceeds what a finite organism can take in.

Your nervous system makes the world livable by selecting, taking an unmanageable torrent and reducing it to a usable stream. That selective coupling is referred to here as filtering. Cell membranes filter, organisms filter, and even a crystal filters when it passes some wavelengths of light while absorbing others. Filtering is the ground floor, not yet consciousness.

Interiority begins to appear when the filtering system starts monitoring its own filtering. A thermostat filters temperature, yet it does not track itself as something with tendencies, blind spots, and vulnerabilities. There is nothing it is like to be a thermostat. A person filters the world and also tracks that filtering: you notice your own reactions, anticipate your own habits, feel when something is off, and sometimes recognize that what is off is your own way of seeing. That recursive turn, filtering that includes a model of the filter itself, marks the beginning of interiority.

The picture proposed here adds one further requirement: stakes. Some outcomes matter more than others for the system's integrity. The system can be harmed, cut off, destabilized, or pushed into a configuration that undermines the organization it actively maintains. Once a system models its own filtering and also registers that some trajectories are better or worse for it, then there is something it is like to be that system.

This is a framework for consciousness in finite embodied life. It does not try to decide what, if anything, lies beyond that condition. It tries to clarify how experience works here and now while the filter is active.

The central claim is an identity claim. Consciousness, in this account, is a specific kind of organized activity: recursive stake-laden self-modeling, lived from the inside. Brain activity is not being treated as something that produces consciousness the way a factory turns out a product; there is no extra ingredient to add, no ghost to sneak back in. It is one process, though it can be described in two ways, as dynamics from the outside and as experience from within.

This also changes how the framework approaches the so-called hard problem of consciousness. The hard problem starts by separating physical processes from subjective experience, then asks how one could ever give rise to the other. The suggestion here is that the separation comes too early. So the real question becomes less "How does mechanism produce experience?" than "What kind of organization is experience when it is lived from the inside?"

One formal tool matters enough to mention before Part II. The Engagement-Openness-Consolidation (EOC) model (Johnson, 2026), summarized in compact form in the EOC Primer at the end of the book, is the book's first explicit dynamical model of change. In plain terms, durable transformation typically seems to require three things at once: the person must show up to the process, old patterns must loosen, and whatever new form appears must stabilize. The EOC model gives those three processes a small dynamical space. Chapter 11 uses that space first as a visible demonstration of qualia as trajectory. Chapter 12 then uses the same landscape to revisit archetypes.

Figure 10.1 gathers the whole framework into one view: signal becomes livable through successive layers of filtering until the same process is lived from the inside as trajectory.

10.2 Four Levels of Filtering

The framework describes four levels, each one carrying forward what came before. F1 is selective coupling: a crystal interacts with some parts of the signal and not others, filtering light on the basis of physical structure. There is no adapting here, no learning, and no self-model of any kind.

F2 is adaptive coupling. A cell changes how it responds in light of what it has encountered. Synapses strengthen or weaken, immune systems learn, and organisms adjust. At this level, the filter is no longer fixed; what it has encountered begins to shape what it does.

F3 is self-modeling coupling. The system carries an ongoing model of itself as a bounded agent with changing internal states. It can, to some degree, notice its own biases and compensate for them. This is where inwardness first appears, necessary for consciousness in this framework but not yet enough.

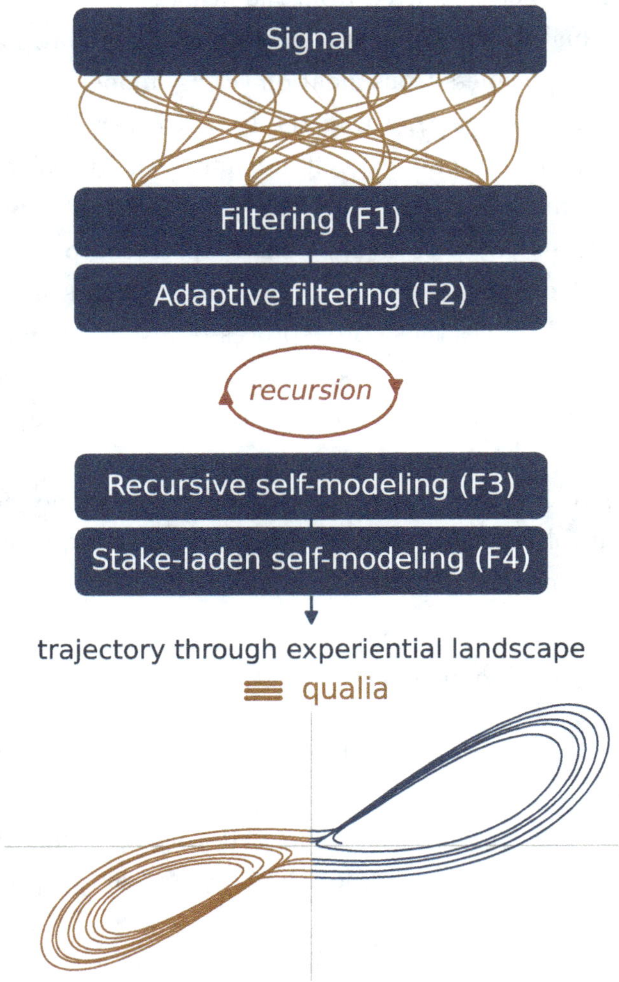

Figure 10.1. The framework in one view. The diagram traces the deepening of a single process, filtering, from signal through selective coupling, adaptive coupling, recursive self-modeling, and stake-laden self-modeling, ending in the lived trajectory through an experiential landscape that this book calls the quale. Conscious experience, in this account, is the lived form of that trajectory in a finite system. Each step marks an additional layer of the same activity, not the introduction of a new ingredient.

F4 is stake-laden self-modeling. The system not only models itself, but some outcomes now matter for its continued integrity, relationships, or values. Consciousness, in the sense defined here, begins when self-modeling becomes stake-laden (F3 recursion with L1 viability stakes). This is where conscious life becomes thick, urgent, and morally weighty.

These are deepening layers of one activity: filtering.

10.3 What Experience Feels Like, Dynamically

Imagine navigating a river in a kayak. You are not contending with the whole river at once but reading the structure of the flow: standing waves, eddies, convergences, and pockets of relative calm. You move by sensing where the water is already inclined to carry you. The skilled paddler does more than react, feeling the pull of the next eddy before it could quite be explained in words.

Inner life has this same sort of structure. Creative discovery, falling in love, breaking a habit, spiraling into obsession, and coming unstuck in grief are trajectories through a dynamical landscape, not different "contents" inside the mind. Some paths are easy to enter and hard to leave. Others constrict your possible actions almost to nothing. Still others widen the field. At times, the whole landscape feels brittle and unstable; at other times, it feels spacious.

That is why the framework uses the language of basins, attractors, separatrices, and transitions. Those terms are a way of saying that experience has shape, not a mathematical gloss laid over psychology. A deep, narrow basin can feel like entrapment, while a broad, shallow basin gives a sense of room to move; a sudden crossing may be felt as a crisis or a breakthrough, and a widening horizon as wonder.

Qualitative feel and dynamical form are two descriptions of the same event. The most important refinement is this: qualia are trajectories through experiential landscapes. The landscape constrains motion. The lived quality is the motion itself. This is why the moving attractor in Chapter 11 matters. A still picture can show the landscape, but only motion shows the kind of object that experience is.

10.4 Why Some Things Matter

Consciousness requires more than self-monitoring. For the one doing the monitoring, something also has to be at stake. The framework describes three broad levels of mattering.

At the most basic level, the system can be harmed. Its boundary can be breached, its energy can be depleted, and its existence can be threatened. This is the level of primal valence: approach and avoidance, comfort and danger.

At the relational level, the system can lose connection. Attachment, rejection, belonging, abandonment, intimacy, and grief live here. These experiences matter because the system maintains itself in relation rather than in isolation.

At the normative level, the system can betray its own commitments. Truth, beauty, justice, loyalty, compassion matter not just because they preserve the body or protect a bond, but because the system has organized itself around principles that define what it is trying to be.

These three levels do not replace one another; they build on each other. A human moral crisis can involve all three at once: bodily threat, relational loss, and the risk of becoming someone one does not want to be.

10.5 A Simple Way to Think About the EOC Model

Because Chapters 11 and 12 lean on the EOC model, it helps to carry one simple picture into them. Engagement (E) means asking whether I am turning toward the process or away from it, whether I am actually showing up or withdrawing. Openness (O) means asking whether the old patterns are loosening enough for anything new to happen or whether the system is clamping down to protect what it already knows. Consolidation (C) asks whether the change is becoming stable, whether a new pattern is really taking hold or vanishing as soon as the moment passes.

The model's central intuition is that real change usually requires all three to occur together. When stability arrives without engagement and openness, it simply entrenches whatever pattern already occupies the person. Human beings consolidate growth, certainly, but we also consolidate defense.

The model is useful here because it is small enough to see. In one regime, it settles. In another, it approaches transition. In another, it moves unpredictably while remaining bounded. That visible motion gives us a concrete example of what qualia-as-trajectory means. A formal version of this picture, with the three differential equations and the regime structure that follows from them, is given in the EOC Primer at the end of the book.

10.6 What Part II Does

Part II takes this framework and puts it to work. Chapter 11, "The Shape of Qualia," uses the EOC system as the minimum visualizable case of a trajectory moving through a structured inner landscape. The chapter does not claim that the whole psyche is three-dimensional, only that qualia have trajectory structure.

Chapter 12 uses that trajectory picture to return to archetypes.

The Shadow, the Hero, the Trickster, the Great Mother, and the Self are treated as recurring landscape conditions, not inherited pictures in a mystical storehouse. The Shadow receives the strongest formal treatment because the model supplies a clear mirror structure.

Chapter 13 extends the filtering hierarchy across deep time, from minerals to cells to forests to silicon, grounded in the author's geological work and in the larger history of Earth systems.

Chapter 14 examines what happens when the filter opens: contemplative practice, psychedelics, dissolution of ordinary boundaries, and the recurring reports of "clear light," ego softening, and reorganization.

Chapter 15 applies the same picture to addiction and relapse, where the language of deep basins, external perturbation, and fragile reconsolidation becomes painfully concrete.

Chapter 16 turns to the carbon-silicon conversation: the AI systems that helped develop the argument and the question of whether the organizational conditions described here might arise outside of biology.

Chapter 17 states what remains unresolved, what would count against the view, and what it leaves open about the nature of the signal itself.

Each chapter ends by asking whether the vocabulary actually helps with difficult cases. That is the point of Part II. The framework has to do work.

Part II

The Framework Applied

CHAPTER 11

The Shape of Qualia

The claim that qualia are trajectories through experiential landscapes is easier to comprehend when you can see them in action. A sentence can say that a feeling has a shape, but a sentence cannot by itself show how the shape moves, how it returns, how it hesitates at a boundary, how it falls back into the old basin, or crosses into another form. For that, we need motion.

For the purposes of this book, dynamical systems theory is not merely a useful tool but the language in which the central claim becomes visible. That is why this chapter turns to the Engagement-Openness-Consolidation (EOC) model developed in Johnson (2026). The model was built to describe psychological transition: how a person moves near thresholds of change, where old organization loosens and new organization may or may not take hold. Here it serves a different purpose: it gives the book a moving landscape in which the claim that qualia are trajectories can be followed visually. The EOC Primer at the end of the book provides the equations and regime structure in summarized form for readers who want the formal backbone alongside this chapter.

A live version of the attractor runs at `innerattractors.com`. The site serves as a companion to this book because the central object of the chapter is not a static diagram. Rather, it is the trajectory orbiting one wing, leaving it, crossing the separatrix, orbiting another wing, returning, and changing character as the driving intensity changes. A static figure can show the outline, but the moving attractor allows the reader to see the dynamical nature of the framework developed here.

11.1 Trajectory, Not Location

Once the framework becomes dynamical, the usual picture of qualia as inner locations becomes too simple. A first cartoon might place anger in one region, joy in another, and grief somewhere else. That picture has some use, especially when the reader is first learning to think spatially about experience. But its usefulness is limited because experience does not sit still.

When anger gathers, it tightens the body, recruits memory, narrows the future, looks for confirming evidence, and then either releases, hardens, or folds into something quieter. Grief comes in waves, often after long intervals in which the person appears to be living elsewhere. Relief is not a coordinate either; it is a sudden change in curvature, a loosening of the field when the feared future loses its grip. Love, in its mature forms, is not just a pleasant state visited from time to time. It can become a durable way of moving through the world, a repeated return toward compassion, trust, repair, and recognition.

Even numbness, which seems at first like the absence of movement, has trajectory structure. The system is not nowhere. It is returning again and again to a narrow range in which little else can enter. That return is part of what numbness feels like from within. The stillness is made and maintained.

The landscape gives experience a structured space to occupy. It determines which movements are easy, which are costly, which regions pull the system back, and which boundaries make small differences matter. The trajectory is the lived event, unfolding through that landscape over time.

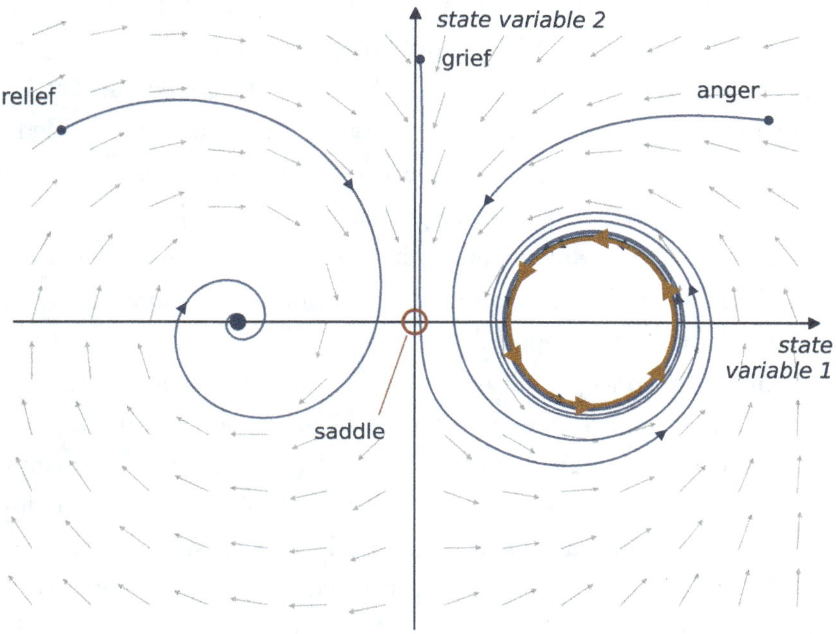

Figure 11.1. A structured experiential landscape with two attractors. The left basin has a stable focus (filled circle); the right basin has a stable limit cycle (bronze loop). The two basins meet at a saddle (open circle) at the origin. Three trajectories show distinct shapes: *relief* spirals inward to the focus; *anger* spirals through several orbits before locking onto the limit cycle; *grief* descends along the saddle's stable manifold, threads through the slow region near the origin, then swings around the lower portion of the right basin before settling onto the cycle from below. In the framework, qualia are recurrent curves of this kind, not labeled positions within the landscape.

Chapter 5 stated the general claim that qualia are trajectories through a structured landscape. This chapter gives the reader a physical, dynamical landscape where time-dependent trajectories can be viewed directly; Figure 11.1 marks the first shift, from a quale as a point to a quale as a path.

This shift has precedent. (Fekete and Edelman, 2011) argued

that experience cannot be identified with an instantaneous activation point, because even silent units help determine what the system can become next. (Moyal et al., 2020) later developed this into Dynamical Emergence Theory, where phenomenal content is tied to the topology of trajectories through collective neural activity spaces. Their vocabulary differs from the one used here, but the underlying concept is the same: from point to path, from snapshot to motion, from isolated activation to the structured unfolding of a system through its own state space.

This same concept will matter later for silicon-based systems. A large neural network does not simply occupy one meaningful vector and then another. As a prompt unfolds, its internal state moves through a high-dimensional activation manifold. Possibilities form and decay; representations are stabilized, revised, suppressed, or carried forward. The EOC system is the reduced case, the one we can see operating in three dimensions of space and one dimension of time. A large neural network is the too-many-dimensions-to-visualize case, with thousands to millions of relevant activation dimensions depending on how the state is represented. But the structural question is the same in both settings: what kind of trajectory is moving through what kind of landscape, and does the system have the recursive self-modeling and stakes required for that trajectory to be lived from within?

11.2 The Minimum Visualizable Case

The framework needs a dynamical tool that can make motion through an experiential landscape visible. Consciousness almost certainly unfolds in a state space far too high-dimensional to draw, so the example has to be simple without becoming trivial. It must still show basins, residence, return, thresholds, separatrices, sensitivity, and changes of regime. The simplest useful tool I have found is the three-variable dynamical system published in 1963 by the meteorologist

Edward Lorenz (Lorenz, 1963). Lorenz developed it as a reduced model of atmospheric convection. Many readers will know its sensitivity to initial conditions through the butterfly effect: in the chaotic regime, a minute difference in the starting state can eventually redirect the entire trajectory. The system's familiar two-winged attractor gives that motion a shape the reader can see.

The atmospheric origin of the equations does not confine them to weather. The same equations were later derived independently for single-mode laser dynamics, even though the variables refer to entirely different physical quantities (Johnson, 2026). What carries across domains is not the material but the coupling structure: the way the variables act on one another. The labels may change, but the relationships among them cannot be assigned arbitrarily. Under the coupling assumptions of the EOC model, Engagement, Openness, and Consolidation obey exactly the Lorenz equations under relabeling (Johnson, 2026). Lorenz used the equations to study atmospheric convection and predictability; EOC proposes the same coupling structure as a conjectural model of psychological transition; and this book uses the resulting landscape more narrowly still, as a way to visualize qualia as trajectories. Its role here is visual and structural, not a claim that consciousness is three-dimensional or that the mind is a weather system.

In the EOC interpretation, one wing carries positive Engagement and positive Openness; the other carries negative Engagement and negative Openness. Consolidation measures the amount of pattern formation, not whether the pattern is healthy, which matters psychologically. A person can consolidate growth, but a person can also consolidate defense. The dynamical system has no sentimentality built into it. It stabilizes patterns when the coupling conditions support pattern formation.

The system also has thresholds. Below the first threshold, perturbations decay back toward the origin. In the ordered regime, two nontrivial fixed points exist, and trajectories can settle into one wing or the other. Above what is called the Hopf threshold, those fixed points lose stability. The trajectory may orbit one lobe for a while, cross to the other, orbit there, and cross again. In chaotic portions of the post-Hopf regime, it remains bounded without becoming predictable in the ordinary point-by-point sense.

The same reduced system can therefore support a return to baseline, stable residence, near-threshold instability, or unsettled motion depending on the driving intensity. Figure 11.2 shows the contrast between stable residence and post-Hopf motion.

11.3 Residence and Return

Start with the ordered regime, because it is the easiest to recognize in ordinary life. A trajectory enters one wing and spirals inward. Released into the other wing, it does the same there. Each wing can hold the system. Each can receive perturbations and still draw the motion back.

Think of a person who is chronically angry with themselves, or with someone else. The anger is not merely a thought that recurs often, but instead has become a route of return. The person may be distracted from it, softened for a moment, amused, comforted, or even moved by generosity, and still the motion eventually curves back. The world is filtered through injury, anticipated insult, and rehearsed justification. Nothing has to be decided each time; the basin simply receives the trajectory.

Now think of a person in a long good marriage, a deep friendship, or a vocation that has become a genuine home. There is return here too. Fatigue, irritation, boredom, and disappointment still occur, but the trajectory bends back toward trust, repair, shared meaning, and renewed participation. The structure is not a passing pleasant mood; it is a durable way of moving through a relational landscape.

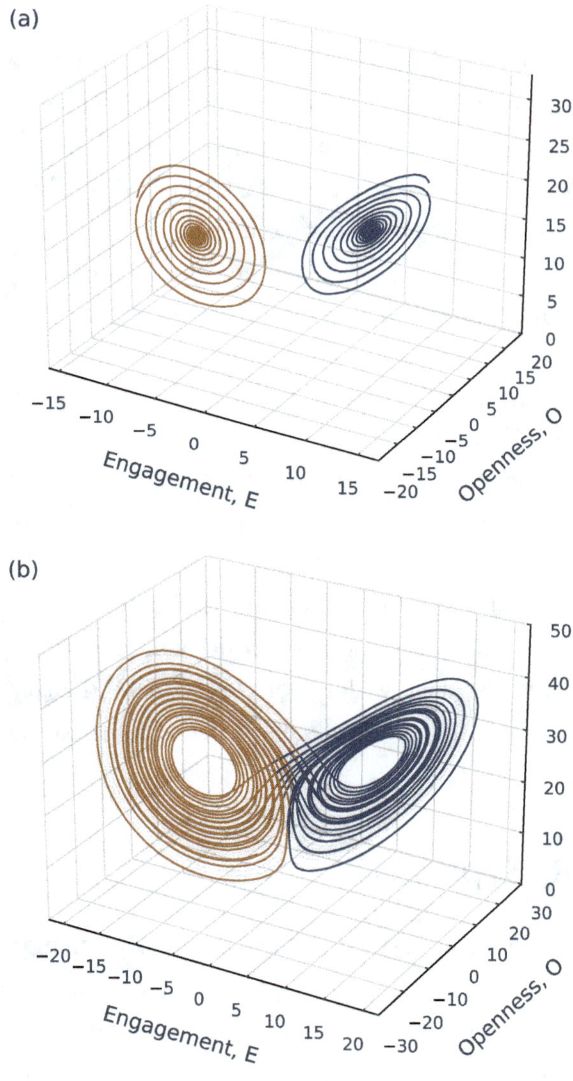

Figure 11.2. The EOC system in two regimes. Panel (a) shows the ordered regime, where driving intensity sits below the Hopf threshold: trajectories spiral into stable foci, one in the growth wing (blue) and one in the defense wing (bronze), and remain there. Panel (b) shows a chaotic post-Hopf example: trajectories no longer settle into either fixed point and instead move through both wings, producing the two-lobed attractor. The contrast is between stable residence and unsettled exploration. Both arise from the same system; only the driving intensity differs.

These cases differ morally and emotionally, but their landscapes can be compared. Each involves residence and return, and each resists perturbation. One consolidates defense, the other growth. One narrows around injury, the other returns toward relation. The signs of the underlying variables carry the content, while the geometry of the basin carries the form of entrenchment.

This helps explain why ordinary emotion words are both useful and blunt. Anger can mean a brief excursion, a tight orbit, a recurring loop, or a life organized around grievance. Joy can mean a sudden opening, a lightness that passes, or the stable return of a person whose life has room for delight. The word points toward a region of experience, but the quality itself depends on the path: how long it remains, how quickly it returns, how hard it is to leave, how much of the future it opens or closes.

A feeling therefore has both content and shape at the same time. The content is what the motion is about, whereas the shape is how the motion holds, returns, widens, or closes. The identity thesis does not choose between them, but says both belong to the organized dynamics.

11.4 The Separatrix

Between the ordered wings lies the boundary separating their basins of attraction. In dynamical systems language, that boundary is a separatrix. The term is technical, but the lived situation is familiar. It is the region where the old pattern has loosened enough to lose inevitability, while the new one has not yet become stable enough to be trusted.

Anyone who has gone through real change knows this region. A person may be leaving a relationship, entering recovery, beginning therapy, losing faith, finding faith, changing work, writing a book, or admitting a truth that has been kept hidden. For a while, the person is not securely one thing or the other. Small events carry disproportionate force: a conversation, a bad night of sleep, an unexpected kindness, a frightened retreat. From the outside, the event may look too small to explain the outcome, but from inside the system, the person is near a boundary.

That is why the separatrix belongs inside a framework for qualia. The wobble before change has its own felt character of experience. It is the felt form of a trajectory moving through a region where nearby futures genuinely diverge, neither confusion nor a mixture of two settled states. The person feels unstable because the system is unstable in just this sense: small differences can decide which basin receives the motion.

The companion site `innerattractors.com` helps here because sensitive dependence is hard to feel from a static picture. Two trajectories can begin almost together and then separate dramatically. A small perturbation near the edge of a basin can matter more than a larger perturbation near its center. That is the structure of the landscape. Figure 11.3 gives the static version of that lesson: two nearby paths, one returning and one crossing.

This gives a more precise way to speak about crisis and breakthrough. The dramatic part of change is often not the size of the push, but the location of the trajectory when the push arrives. A word, an intervention, a loss, or an act of courage can appear to do little for years and then suddenly matter enormously because the system has moved near a crossing.

11.5 Regime and History

The same landscape behaves differently as the driving intensity changes. Below the first threshold, the origin is stable. A perturbation can disturb the system, but the disturbance decays. Push the trajectory a little and it returns; push it again and it returns again. No durable wing forms because the landscape has not crossed the threshold at which sustained reorganization becomes possible.

The felt form is return without consolidation. A person may have the right conversation, make the right plan, recognize the right pattern, and still feel the whole thing fade by morning. Something moved, but nothing took. The experience can be discouraging because it is not blankness. There is enough motion to suggest possibility, but not enough driving intensity for the motion to become a new form.

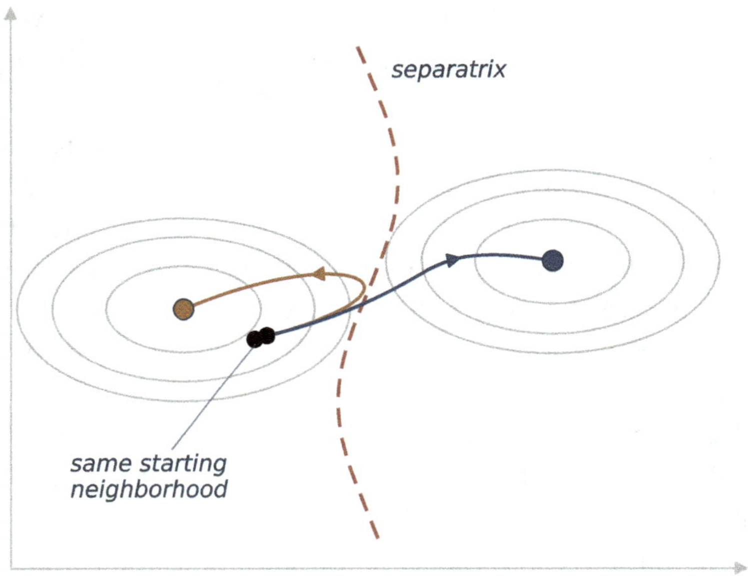

Figure 11.3. The separatrix between basins of attraction in the EOC landscape. Two nearly identical trajectories approach the same boundary; one returns to its original basin while the other crosses into the complementary one. The figure makes visible a structural fact that runs alongside a familiar lived one: small differences can decide whether a person returns to a prior pattern or crosses into another. Near a separatrix, the location of the trajectory when a perturbation arrives can matter as much as the size of the push.

In the ordered regime, pattern can settle. The trajectory enters a basin and returns there after disturbance. The settled pattern may be adaptive or defensive. The model is useful partly because it refuses the comforting idea that stability always means health. Human beings consolidate safety, skill, love, and courage; we also consolidate resentment, avoidance, shame, and fear.

Above a specific driving intensity, the Hopf threshold, what were once stable fixed points now lose stability. In the post-Hopf region, the trajectory can orbit one lobe, cross, orbit the other, and cross again without settling into either. The motion may be aperiodic, transiently complex, or caught in a periodic window, so the cautious claim is loss of stable fixed-point residence, not featureless chaos at every value of the driving parameter.

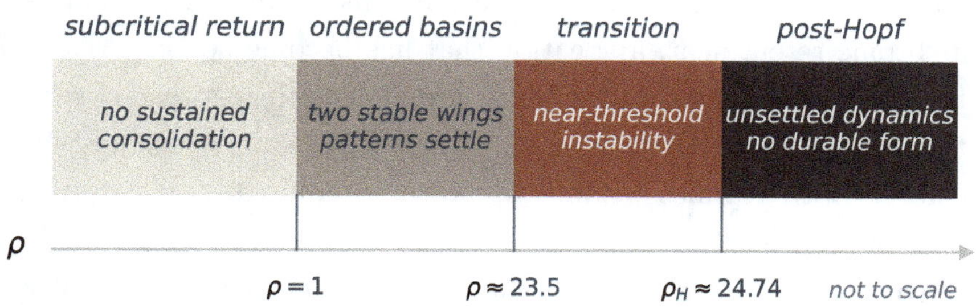

Figure 11.4. Regime structure of the EOC system along the driving intensity ρ. Below the first threshold, perturbations decay back to baseline and the system shows return without consolidation. As ρ increases, two stable foci emerge and the trajectory can settle into the growth wing or the defense wing. Near the Hopf threshold, stability weakens. Beyond it, stable fixed-point residence is lost, and the post-Hopf regime can include aperiodic wing-switching, transient complexity, and periodic windows. The same landscape therefore supports several lived possibilities: return to baseline, stable residence, near-threshold instability, and post-Hopf exploration, depending on how hard the system is being driven.

This gives us a picture of crisis without consolidation: opening, closing, beginning again, retreating, insight appearing and failing to hold, defense returning and failing to protect. The system has energy, sometimes enormous energy, but the motion does not become a durable form. In the chaotic portions of the post-Hopf regime, the path has order without repetition. It seems almost ready to settle, and then it leaves. It circles one wing long enough to suggest commitment, then crosses to the other. The broader lesson is useful even when the dynamics are not chaotic in the strict sense: a pattern can be intelligible without being point-predictable. A person can be deeply committed to their patterns and still be impossible to predict in detail from one moment to the next. Figure 11.4 places these possibilities along the driving parameter, from return without consolidation to ordered residence and post-Hopf exploration.

The ordered regime is less visually arresting but structurally more common in ordinary life. Trajectories settle into one wing, perturbations return to it, and durable patterns form: a long marriage, a chronic resentment, a vocation that has become home. Crisis and breakthrough in the ordinary sense are then often best read as separatrix crossings within the ordered regime, not as entries into post-Hopf chaos. Both regimes belong to the same landscape, distinguished only by how hard the system is being driven.

History then adds another layer. In a nonlinear system, where the trajectory ends up depends on the path it took, not just on the current parameter value. A parameter can return to an earlier value while the trajectory keeps its new organization. In some systems this is hysteresis: the route in and the route back are not the same.

The broader point is path dependence, and the lived version is everywhere. A person enters crisis under one set of pressures. Later, when some of those pressures lift, the person does not simply go back. Recovery is not rewind. Grief does not resolve by subtracting the cause. Addiction recovery does not erase the basin of addiction cut into the landscape. Broken trust, even when repaired, does not leave the relationship identical to what it was before. The system carries history in its current geometry.

This is another reason why the point picture fails: the quality of a state depends partly on the path that brought the system there. The same apparent calm can be exhaustion, peace, resignation, or trust, depending on the trajectory that led into it and the futures available from it. A point by itself has no history; a trajectory does.

11.6 Toward a Phenomenometer

The EOC attractor has done its first job once qualia can be seen as motion through a landscape. The next question is whether the motion can be measured. A structural geologist who wonders what stress was required to cause a specific type of rock deformation does not see the stress directly. It is more of a detective game. Stress leaves signatures behind in mineral grain boundaries, dislocation structures, crystallographic fabrics, and faulted and folded layers, all of which carry the history of the forces that acted on a rock. With calibration, this evidence becomes measurement. The same ambition belongs here in relation to qualia. If qualia are trajectories through the landscape, then features of trajectories should map meaningfully onto features of felt life.

Recall the phenomenometer introduced in Chapter 5 (Table 5.1): basin residence, return rate, escape cost, separatrix proximity, sensitivity, control centralization, action-space compression, accessible future diversity, and velocity and curvature. Each was named there as a candidate coordinate, defined informally enough to invite intuition without yet committing to numbers.

With the EOC system now in hand, several of these coordinates can be tied directly to quantities computable from the simulated trajectory: residence time on each wing, relaxation rate after perturbation, and so on. The remaining coordinates would require domain-specific mappings from trajectory geometry to control allocation, action space, report, physiology, or behavior. Naming has acquired the structure that measurement would require, even though calibration remains future work. The next subsection asks whether that structure can actually distinguish felt states. Figure 11.5 gives this next step a visual form before the prose walks through it.

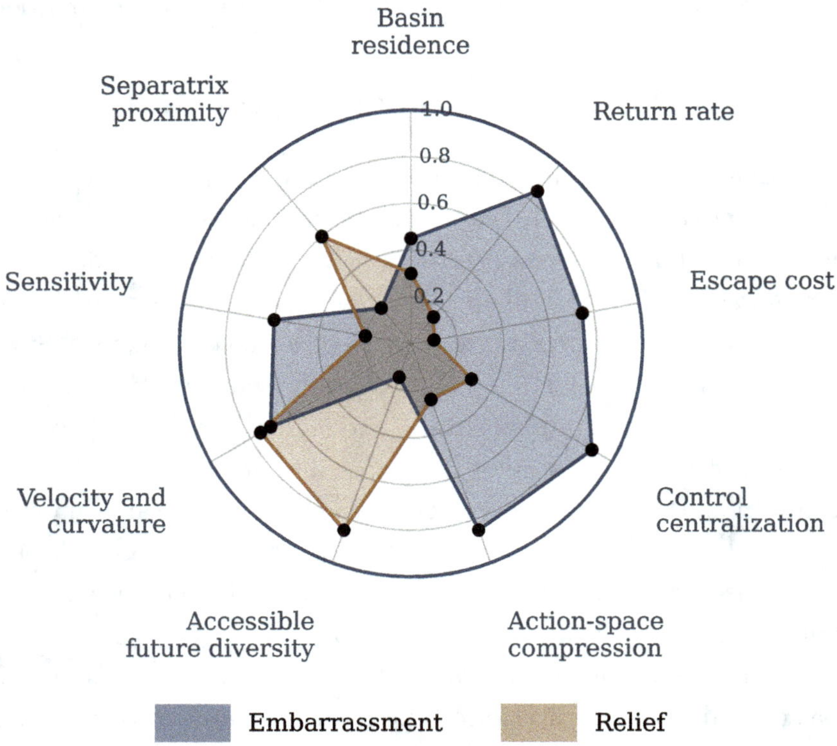

Figure 11.5. Two illustrative phenomenometer signatures. Embarrassment (blue) shows high values on the four axes that mark constraint (Return rate, Escape cost, Control centralization, and Action-space compression), with Accessible future diversity collapsed. Relief (bronze) shows the inverse pattern, with Accessible future diversity high and Separatrix proximity lifted by the recent crossing out of the threat basin. The two shapes are visually distinct at a glance, supporting the framework's claim that these two states are different *shapes* of trajectory, not different points in an inner space. Both states score similarly on Velocity and curvature, because the radar plot represents only the magnitude of motion; the direction of curvature differs (sharp inward for embarrassment, sharp outward for relief) and is described in the body text. Values are illustrative levels on a 0–1 scale, not fitted measurements.

Two Worked Signatures: Embarrassment and Relief

A table of candidate coordinates states, in effect, that nine features should help discriminate one quale from another. To see whether the proposal is more than vocabulary, the test is to take two familiar experiences and ask whether the coordinates actually identify a structural difference between them. Embarrassment and sudden relief are good test cases. Both were introduced in Chapter 5 as the opening phenomenological anchors of the qualia chapter. Both involve sharp transitions in the self-model. Both are within ordinary experience for nearly every reader. And, as Chapter 5 noted, both are structured in ways that complicate the simple picture of pleasure and pain. They differ from one another in ways the phenomenometer should be able to register.

Take the dinner-party embarrassment first, in the form introduced in Chapter 5. You make a joke that lands badly. Your face warms. The conversation continues, but for a few seconds you are no longer fully in it. You are caught in an urgent self-monitoring loop, seeing yourself as you imagine the others see you. The available moves narrow all at once.

Walk through the phenomenometer coordinates one by one, in the order of Table 5.1. Basin residence: short to moderate. The state holds for seconds, perhaps a minute, rather than being the kind of thing that persists for hours; though it can be reactivated later when the memory returns. Return rate: high. A quick shift of attention does not move the state; the trajectory keeps spiraling back to the imagined social judgment. Escape cost: relatively high in the moment. The state does not yield to ordinary attempts to redirect attention. Separatrix proximity: low. The trajectory is deep inside a recursive social-monitoring basin and not near a transition boundary; the state will most likely terminate when the conversation moves on, not when the system reorganizes from within. Sensitivity: moderate to high. Small additional perturbations, like an awkward pause from another person, can deepen the state quickly. Control centralization: high. The imagined social judgment dominates regulation; other concerns, however urgent in principle, recede until the state passes. Action-space compression: high. Few responses feel available; many ordinary moves, like making another joke, suddenly feel impossible. Accessible future diversity: collapsed. The next several seconds feel constrained to options about how to recover, hide, deflect, explain. Velocity and curvature: a sharp inward bend at the moment of the social misfire, then a tight orbit around the imagined view of oneself.

Now take sudden relief in the form introduced in Chapter 5. You think you have lost your wallet, or forgotten an important email, or made a mistake that will be hard to undo. The body tightens around that possibility. The trajectory begins to curve into threat. Then you check the other pocket, or open the sent folder, or notice that the feared outcome was not real. The shift is immediate. Breathing deepens. The grip loosens. More than one action becomes possible again.

Walk through the same nine coordinates. Basin residence: very short for the relief moment itself, though the post-relief calm can extend into something longer-lasting. Return rate: low. The threat trajectory does not snap back; the system has just been released from it. Escape cost: very low; in fact, the escape has just happened. The relief is the felt aspect of the system having already left the threat basin. Separatrix proximity: the trajectory has just crossed and is now receding from the boundary. Sensitivity: low in the moment. Small perturbations do not easily push the system back into threat once the threatened outcome has been ruled out. Control centralization: falling. Where the threat had captured regulation around a single anticipated outcome, attention can now distribute back across the ordinary field. Action-space compression: low and falling. Many actions feel possible again; the constraint that organized the previous moment has lifted. Accessible future diversity: rapidly expanding. Where seconds before the future had narrowed around managing a feared outcome, now it widens to include the ordinary range of possibilities. Velocity and curvature: a sharp outward release, the curvature of an opening rather than a closing.

Table 11.1. Illustrative phenomenometer signatures for two familiar experiences. The values are qualitative levels, not measurements. The point is structural: the proposed coordinates discriminate two experiences along several dimensions in ways that match their felt differences.

Feature	Embarrassment	Sudden relief
Basin residence	Short to moderate; reactivatable	Brief release moment, longer post-state
Return rate	High; trajectory returns to the loop	Low; the threat trajectory does not snap back
Escape cost	High in the moment	Negligible; the escape has just occurred
Separatrix proximity	Low; trajectory deep in basin, far from boundary	Recently crossed; trajectory now receding
Sensitivity	Moderate to high	Low; threat reactivation is unlikely
Control centralization	High; imagined social judgment dominates regulation	Falling; attention redistributes once the threat is released
Action-space compression	High; many ordinary moves feel impossible	Low and falling; many actions feel possible
Accessible future diversity	Collapsed around recovery options	Rapidly expanding back to ordinary range
Velocity and curvature	Sharp inward bend; tight orbit	Sharp outward release; opening curvature

The two signatures differ along nearly every coordinate, and the differences correspond to the felt difference between the two experiences. Embarrassment narrows and holds, moving the trajectory deeper into a basin; relief opens and releases, moving the trajectory out of one. In this illustrative case, the phenomenometer coordinates do discriminating work, not merely supplying a vocabulary for the experience after the fact.

Two cautions matter here. First, the entries in Table 11.1 are illustrative levels, not numerical measurements. They have not been fitted to time-series data because the relevant time-series data have not been collected. The point of the table is structural, not quantitative: it shows that the phenomenometer coordinates, taken together, distinguish two states in ways that correspond to their felt differences.

Second, neither embarrassment nor sudden relief is a single uniform experience. The values would shift across persons, contexts, and developmental histories. A chronically shame-prone person would show different residence and return values for embarrassment than someone who recovers quickly. A person whose threat baseline is already high would show different relief signatures than someone whose baseline is calm. The framework expects these differences. It does not pretend to flatten them into a single canonical reading. What the worked signatures show is modest but important: the phenomenometer is more than a list of plausible-sounding dimensions. It is a structured way of asking what makes one quale different from another. That is the demonstration the table by itself could not provide.

The methods needed for this are already being developed. Garland, Bradley, and Meiss showed that the topology of dynamical reconstructions can be explored from time series, including the Lorenz system (Garland et al., 2016). Moyal, Fekete, and Edelman argued that geometric measures of trajectories are the right route into phenomenal structure (Moyal et al., 2020). The EOC model adds an explicit, reduced psychological state space in which the larger phenomena can be made visible (Johnson, 2026). The hard work is calibration. For that, Engagement, Openness, and Consolidation would have to be measured in real time series. Landscapes would have to be estimated for persons, dyads, groups, or silicon-based systems, depending on the domain. Basin residence, return rate, escape cost, separatrix proximity, sensitivity, and action-space compression would have to be compared against first-person report, behavior, physiology, and response to intervention (Table 5.1). A phenomenometer would not be a single dial marked consciousness. It would be a way of reading the geometry of lived motion.

11.7 What the Example Shows

The Lorenz/EOC system gives the identity thesis a visible center. It shows how a simplified landscape can hold residence, return, crossing, threshold, sensitivity, path dependence, and regime change in one moving object. Those are not decorative features added to feeling from the outside, but are among the features that give feeling its shape.

Anger, grief, love, fear, relief, and wonder are therefore better understood as families of trajectories than as fixed inner locations. A feeling can be determinate without being predictable. It can occupy a region and still depend on the route by which it arrived. It can return with such force that insight cannot stop it. It can cross under a perturbation that looks small if the observer cannot see the boundary.

This chapter also gives the rest of Part II a hinge. The archetypes chapter coming next can treat recurring figures as recurring circumstances shaping the landscape. The addiction chapter can speak of basin depth, residence, and return rate without leaving the language of experience. The altered-states chapter can describe filter opening as a change in the landscape through which trajectories move. The AI chapter can ask whether high-dimensional silicon-based systems trace internally structured trajectories of the relevant kind.

The compact statement is this: the landscape is the space of possible experience, and trajectory is the lived motion itself. In a recursively self-modeling system with something at stake, that is the shape of a quale.

Chapter 12

Archetypes as Attractor Landscape

Everyone has felt the pull of a new pattern before they could name it. An idea keeps returning until it feels like a demand rather than a thought. A relationship begins rearranging your priorities before you have consciously decided anything. A conversation leaves the unmistakable sense that an older version of yourself has just been challenged by a newer one. You have not changed yet, but the pull is unmistakable.

Depth psychology gives names to these pulls. Jung used *archetype* for inherited organizing forms whose symbolic expressions recur across cultures, centuries, and individuals who have never encountered one another (Jung, 1959). The Shadow, the part of yourself that you cannot yet see as your own. The Hero, the pressure to leave one way of being before the next one is secure. The Great Mother, the container that holds a fragile system while it develops. The Self, the pattern that can organize the others without flattening their differences.

For more than a century, the status of archetypes has remained unsettled. Are they inherited images in a collective unconscious? Cultural templates passed down through story and ritual? Merely vivid metaphors for common human situations? The debate has lasted because the available language has usually forced a bad choice between hidden inner images and empty cultural residue. Within this framework, archetypes are best treated as recurring trajectory forms in complex filtering systems that have recognizably human stakes.

Chapter 11 shifted the ground beneath this question, arguing that qualia are trajectories through the landscape, and it used the Lorenz/EOC system as the minimum visualizable case. This implies that archetypes are recurring forms of movement through a landscape: returns, crossings, projections, holding patterns, and rhythms of integration. The Shadow provides the strongest case. It is not merely analogous to an attractor; in the EOC landscape, it finds a formal analog in the mirror structure itself. Once this is established, the discussion can widen to the other archetypes, where the mappings are more suggestive.

12.1 From Analogy to Dynamical Form

Ernest Rossi (1989) and John Van Eenwyk (1991, 1997) were early proponents of reading archetypes through strange-attractor dynamics, Van Eenwyk in a sustained treatment of the correspondence and Rossi around psychotherapy and change. The language of nonlinear dynamics really does sit surprisingly close to the language of depth psychology. Bifurcation sounds like a crisis. Stable and unstable landscapes sound like habit and transformation. Strange attractors sound like the persistent patterns into which inner life keeps settling.

Analogy helped Rossi and Van Eenwyk find the neighborhood, but not yet the machinery. In the works cited here, they could say that archetypes were like attractors, but not which variables, parameters, or equations would generate them, or what evidence would count against the proposal. That is the line between a suggestive metaphor and a scientific claim.

The EOC model crosses that line far enough to make the question more explicit. It was not built as an independent proof of Jungian categories, but it allows us to ask a more exact question: if a self-organizing psychological system has this landscape, what recurring patterns follow from it?

12.2 What Chapter 11 Established

The previous chapter introduced the Lorenz/EOC landscape as a visible demonstration of qualia-as-trajectory. For the present chapter, only a small portion of that result is needed.

The EOC model has three variables: Engagement, Openness, and Consolidation. Engagement tracks approach and withdrawal. Openness tracks plasticity and rigidity. Consolidation tracks accumulated pattern formation. Engagement and Openness are signed process variables. Consolidation is directionally neutral: a person can consolidate growth as well as defense. The full architecture, including the equations, parameters, and bifurcation structure that produce the wing geometry, is given in the EOC Primer at the end of the book; this chapter draws only on the four-quadrant structure generated by those dynamics.

In the ordered regime, the system has two nontrivial wings. One is the growth wing: positive Engagement, positive Openness, and adaptive consolidation. The other is the defense wing: negative Engagement, negative Openness, and defensive consolidation. Both are genuine organizations. Figure 12.1 compresses that result into the Engagement-Openness plane: the stable wings lie on the diagonal, while the off-diagonal quadrants are transitional, not self-sustaining.

That point matters because archetypes are not static labels applied to experience from the outside but are instead recurring modes of motion. The growth wing and defense wing are regions into which trajectories enter, orbit, return, and from which they may or may not escape, not places where a person simply sits. A life organized around resentment is a pattern of return. A life organized around trust is also a pattern of return. The moral circumstances differ, but the dynamical fact is shared.

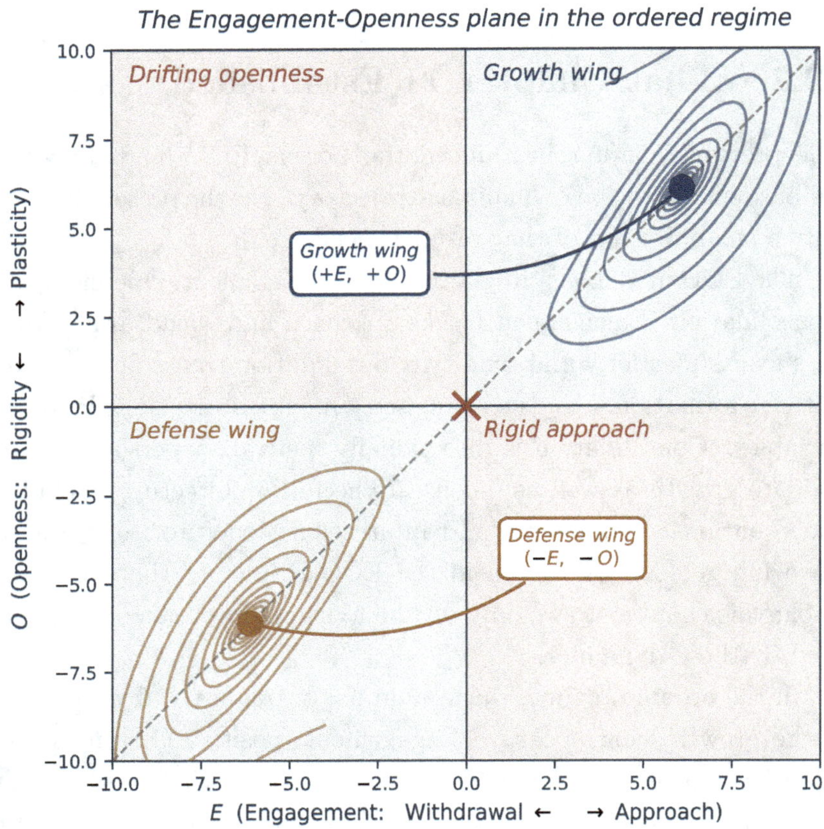

Figure 12.1. The four sign regions of the Engagement-Openness plane in the ordered regime. The two stable wings sit on the $E = O$ diagonal: the growth wing at $(+E, +O)$ and the defense wing at $(-E, -O)$, each surrounded by trajectories spiraling into its fixed point. The off-diagonal quadrants name the unstable regions where engagement and openness carry mismatched signs. *Drifting openness* at $(-E, +O)$ is opening without engaging; *Rigid approach* at $(+E, -O)$ is engaging without opening. Neither off-diagonal region supports a stable basin, and trajectories pass through them only on the way to or from the wings. The origin, marked with an x, is itself an unstable fixed point for $\rho > 1$. Standard parameters and full bifurcation analysis are given in Johnson (2026); the quadrant structure is summarized in the EOC Primer.

The EOC landscape also has a boundary structure. Moving from one wing to the other requires crossing a transition region where small perturbations can matter. This is where ambivalence, crisis, breakthrough, and reversal become psychologically recognizable. The separatrix is the dynamical structure of transition. With those pieces in place, the archetypal language becomes more precise.

12.3 The Shadow as Mirror Complement

The Shadow is often the version of yourself that keeps appearing to you in other people first. Depth psychology names it as the part of yourself you cannot yet own. Chapter 4 placed the Shadow across the stakes hierarchy on a structural footing of its own, grounded in the fact that every formed basin excludes an outside. The EOC model puts that intuition into precise geometric form.

Two moves are at work here, and it helps to keep them separate: the geometry forces the existence of a structurally excluded complement, and identifying that complement with what Carl Jung named the Shadow is then an interpretive mapping, not a further deduction.

At the heart of the model is a simple mirror relation:

$$(E, O, C) \longmapsto (-E, -O, C).$$

In plain language: whatever the person is in the Engagement-Openness plane, there exists a complementary configuration defined by reversing approach into withdrawal and openness into closure while leaving consolidation at the same level. If one wing is occupied, the other does not vanish, but remains structurally available as the complement of the occupied trajectory. That is the Shadow-like structure the model can actually name.

Suppose the system is moving in the growth wing. It is organized around approach, openness, and adaptive consolidation. The complementary configuration, withdrawal, closure, defensive consolidation, remains present in the landscape, but it is not the trajectory the self is currently living. The self, simply by being one thing, excludes its complement. Reverse the example and the logic is the same. A person organized around defense does not eliminate growth. Growth remains present as the live but unoccupied opposite.

This allows us to say several things. First, the Shadow is structured, not random. It is not a miscellaneous pile of whatever the ego happens not to like. In this model, it is the complement of the configuration the self currently occupies. Whatever the self affirms, the Shadow is what that affirmation requires it to deny in order to remain coherent.

Second, projection becomes easier to understand. If the excluded complement remains structurally live inside the system, it is easy to encounter it first as other. The system notices in other people what it cannot yet integrate in itself. Projection, in this view, is a dynamical consequence of occupying one wing of a mirror-structured system.

Third, Shadow integration becomes more exact. It is less a matter of becoming friendly with "the dark side" in the vague therapeutic sense than of changing the system's relationship to its excluded complement. The mirror image no longer needs to remain wholly classified as not-me. The walls between configurations become more permeable, the system becomes less rigidly exclusive, and it becomes more capable of complexity.

Fourth, the Shadow cannot be abolished. As long as a self is organized around one definite trajectory, something complementary remains excluded. A different personality will produce a different Shadow, but it will not eliminate Shadow as such. A self without exclusion would be a self without structure.

That last point matters beyond psychology. Many spiritual traditions imagine liberation as a state beyond division. The model suggests that if organized selfhood remains, exclusion remains with it. The only condition without Shadow would be a condition without organized selfhood at all, a point Chapter 14 revisits from the side of altered states and contemplative dissolution.

The chapter's strongest claim is therefore limited but real: any system with the mirror landscape the EOC model describes generates a complementary configuration that the occupied trajectory must structurally exclude. That structural complement formalizes one piece of what Jung was pointing toward when he described the Shadow.

12.4 Projection, Integration, and the Cost of Becoming Whole

The language of Shadow integration can sound too clean because, in lived experience, integration is usually unstable. A person does not simply discover an excluded complement and calmly add it to the self-model. The old trajectory resists because it has residence, return rate, and basin depth. It is not just an idea, but a way of moving.

That is why projection is so durable; it protects the current organization by keeping the complement outside. If my anger belongs to you, then my self-model remains intact. If your freedom is irresponsibility, then I do not need to feel the rigidity of my own life. If your confidence is arrogance, then my own disowned power can remain safely unclaimed. The projected object carries the landscape the self cannot yet live.

Integration begins when the system can hold enough instability to let the projected complement re-enter. The process is not always pleasant. Sometimes it feels like humiliation, grief, or a strange release. The current self-model has to admit that part of what it has treated as outside was one of its own excluded possibilities.

This is where Chapter 11 matters again. The feeling of integration is a trajectory near a boundary, not the addition of a new belief. It is a passage through a region where the old basin no longer fully holds, and the new organization has not yet stabilized. That is why genuine Shadow work can be so disorienting. The self is changing the landscape by which the model maintains itself.

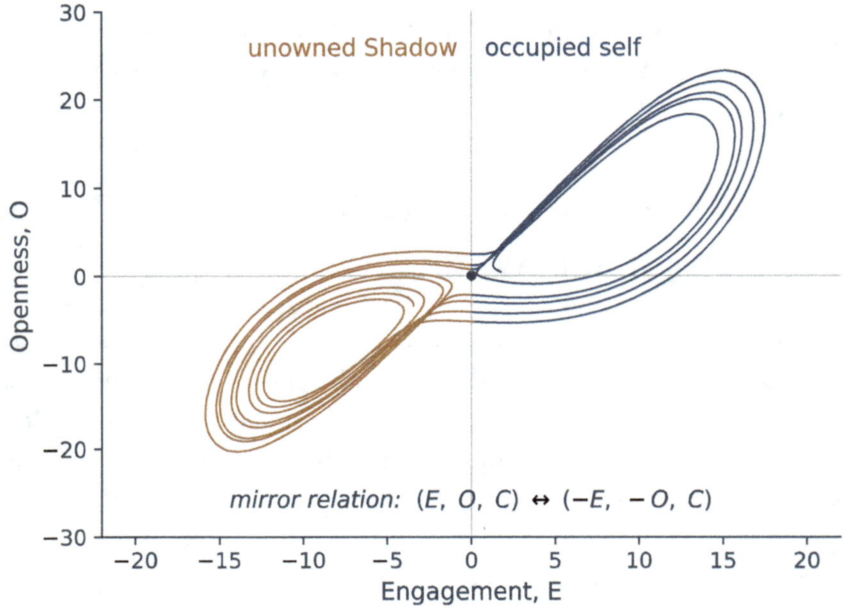

Figure 12.2. The Shadow as mirror structure in the EOC landscape. The conscious self is shown organized in one wing of the attractor; the complementary configuration, defined by the mirror relation $(E, O, C) \mapsto (-E, -O, C)$, sits in the opposite wing as an unowned but structurally live basin. The figure carries the chapter's strongest claim: the Shadow is the unowned mirror basin that the system's own partitioning makes necessary, so a self organized around growth still carries a defense complement and a self organized around defense still carries a growth complement.

12.5 The Other Archetypal Trajectories

The Shadow is the strongest result in this chapter because the symmetry constrains it. The other archetypal forms are not derived with the same formal force; they are proposed as recurring trajectories under common stake structures. What follows should therefore be read as suggestive and testable, not fully derived. Before turning to those less constrained mappings, Figure 12.2 illustrates the Shadow result visually.

Defense and Growth

The two wings of the model are the simplest recurring archetypal forms. One trajectory moves toward novelty and reorganization, whereas the other contracts toward safety and stabilization. Any system with stakes and a landscape like this should generate some equivalent of defense and growth. All that is required is a self-organizing system with something to preserve and something to risk.

Defense deserves respect here. It is often treated as pathology by default, but every living system needs ways to protect its integrity. The problem is not defense as such, but capture behavior: a trajectory that returns to defense even when the world no longer requires it. Growth has a parallel danger. Openness can become destabilization if nothing consolidates. A person can be endlessly available to transformation and never become anything stable. The growth wing is healthy only when movement and consolidation remain coupled.

The Trickster

The Trickster belongs near the boundary. Close to a separatrix, small fluctuations matter disproportionately. The system can surprise itself. A joke can escape before it was intended. A supposedly settled identity says or does the thing it had been refusing to admit. Humor belongs here, and so does irony. So does the embarrassing accident that reveals the truth more plainly than confession would have. The Trickster is what boundary instability feels like when the existing organization can no longer perfectly police itself, not an outside gremlin inserted into the psyche. It prevents total petrification by allowing leakage across the walls the self has built.

The Hero

The Hero is the felt experience of leaving a known organization for a more difficult and more capacious one. In dynamical language, that means crossing the boundary out of a settled basin under real uncertainty. A transition is heroic not because it is large, but because the outcome is not guaranteed. The system may cross and grow. It may fall back or destabilize. The risk is part of the form. This is why heroic stories so often begin with a call that cannot be comfortably assimilated. The call lowers the authority of the old basin, but does not yet create the new one. The crossing is therefore a trajectory through instability, not a march from one finished identity to another.

The Great Mother

The Great Mother is harder to capture in a one-person model because she is fundamentally relational. She is the stable container in which another system can open without coming apart, more than mere nurturance. One system helps hold the basin while the other explores within it. This is why the loss of a containing relationship can be so destabilizing. The person does not only experience sadness, but a basin in which part of their development was being held has partly vanished. The therapeutic alliance, a secure attachment, a durable community, even a trusted landscape can all function in this way. They are not simply contents of experience, but they help shape the landscape through which experience moves.

The Self

The most tempting mistake is to identify the Self with one especially refined fixed point, which would turn individuation into a better kind of entrenchment. A richer possibility is that the Self corresponds to a natural rhythm. The mature system can move between openness and closure, vulnerability and defense, solitude and relation, without being captured by either pole. Greater integration may therefore look less like monotonic consistency and more like organized oscillation: a stable capacity to move between opposites without fragmentation. The Self is not the final place where motion stops, but the deeper organization that lets motion become coherent.

These mappings are more tentative than the Shadow analysis, but they are not arbitrary. They bring archetypal language into contact with a model that can, in principle, be argued with and tested. Figure 12.3 shows two such forms, not as finished doctrines, but as trajectories through the EOC landscape.

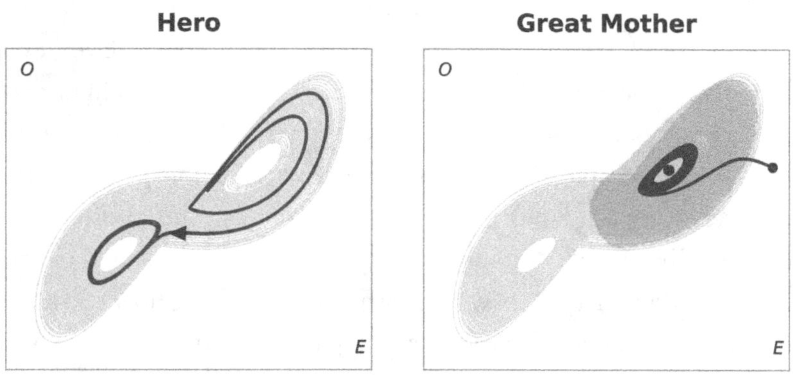

Figure 12.3. Two archetypal trajectory forms in the EOC landscape, projected onto the (E, O) plane. The faint gray traces in both panels show the chaotic attractor at $\rho = 28$ as visual reference for the wing structure; the archetypal trajectories themselves are integrated at lower driving intensities. *Hero* (left): a single trajectory at $\rho = 25$, just past the Hopf threshold ($\rho_H \approx 24.74$), crosses from one settled wing toward its complement; the arrowhead marks the bridging moment. The Hero sits at the edge of stability, not in deep chaos. *Great Mother* (right): the bronze region marks the growth wing's extent, computed as the spatial region the chaotic attractor occupies when in growth mode. The slate trajectory at $\rho = 22$, in the sub-Hopf ordered regime, enters the wing from its outer edge, then spirals clockwise within the basin while converging toward the central fixed point.

12.6 Why Archetypes Recur

Why do the same archetypal patterns recur across cultures, eras, and individuals who have never encountered one another's symbolic systems? The cleanest answer, at least at the level this chapter addresses, is structural. Similar filters navigating similarly structured signals tend to generate similar landscape conditions. Human beings share a great deal: roughly the same bodily architecture, the same mortality, the same developmental vulnerability, the same dependence on attachment, the same exposure to hierarchy, sexuality, danger, grief, cooperation, and loss. We do not live in identical worlds, but we live in worlds constrained in overlapping ways.

Convergent evolution provides a useful analogy. Eyes evolved many times because light is there to be tracked. Wings evolved many times because air and gravity reward lift. Similar problems plus enough complexity produce recurrent solutions, and myth, ritual, literature, and cultural exchange then clothe those solutions in culturally specific images. One culture imagines a dark forest, another a devouring mother, another a dangerous lover, another a desert trial. The symbols differ, but the pressures they answer are often the same.

That is why the framework constrains symbolic interpretation without abolishing it. If archetypal symbols are doing real psychic work, they should cluster around recurrent landscape conditions in human life. This does not settle larger metaphysical readings of archetype, but simply says that recurrence follows naturally from shared constraints.

This also gives the argument a predictive edge. If archetypes are structural, not culturally inherited images, then analogous patterns should eventually appear in sufficiently complex, stake-bearing nonhuman systems facing analogous pressures. For example, one would expect equivalents of complementarity, projection, defensive stabilization, risky transition, containment, and integrative rhythm.

The aim of this chapter is not to prove that every archetypal claim in depth psychology is true, but to show that some of the most durable claims in depth psychology may be structurally recurrent. If that is right, then Jung and his successors were naming regularities before the mathematics was ready for them.

12.7 The Stress Test

The strongest objection is obvious: perhaps this is only a clever relabeling. Perhaps the model is being used to dress old Jungian categories in mathematical language without adding anything real.

The answer is that the chapter earns its keep only where the landscape constrains the interpretation. The Shadow case passes that test most clearly because the mirror relation is already in the model, even though the model was not built to prove Jung right. A self occupying one side of a mirror-structured landscape necessarily excludes a complement. Projection and integration then become consequences of how such a system maintains itself.

The Hero, Trickster, Great Mother, and Self should be treated as proposed correspondences, not derivations. Their value will depend on whether they help organize actual phenomenology, clinical observation, cultural recurrence, and eventually time-series data. The framework becomes stronger where it can say what pattern should appear, under what conditions, and how the claim could fail.

Archetypes do not need to be hidden images in a collective warehouse in order to be real. They can be recurring forms of movement through a shared human landscape, and the trajectories belong to a much longer story. Chapter 13 follows that story back into deep time.

Chapter 13

The Deep Time Signal

Outside the window of my office, white pines stand sixty feet tall, with spruce, hemlock, and maple filling the understory. The most ancient of these creatures may be a century old, but the form they express is vastly older. Trees have been reshaping the planet for roughly 390 million years, an environmental revolution that gained momentum when Late Devonian plants such as Archaeopteris lifted durable canopies above the ground (Algeo and Scheckler, 1998). With forests, the world available to life became very different.

Chapters 11 and 12 asked what attractor landscapes look like from the inside: first as trajectories of feeling, then as recurring archetypal forms. Here, the camera pulls back. If consciousness is what filtering feels like from the inside once the filter has become deep enough to model itself and to have something at stake, then its prehistory is the much longer history of filtering itself. The hierarchy introduced in Part I (F1 through F4) implies that nervous systems intensified and complicated filtering without inventing it. Matter filters, life does so in a more adaptive way, and nervous systems filter through self-models. Across deep time, the same verb reappears under new constraints.

This chapter illustrates how lower levels make later levels possible, without pretending that a pine tree and a person differ only by degree or that the earlier levels were somehow aiming at the later ones.

13.1 F1: The Mineral Filter

The simplest filter is a crystal.

Quartz is silicon dioxide arranged in a trigonal lattice. When light enters quartz, the crystal does not respond equally to everything that arrives. Some wavelengths pass, some are delayed or refracted, and the details of the interaction are determined by lattice geometry. Nothing here is chosen, and nothing is learned. But something crucial is already present: structured selectivity. Figure 13.1 shows the point in its simplest material form.

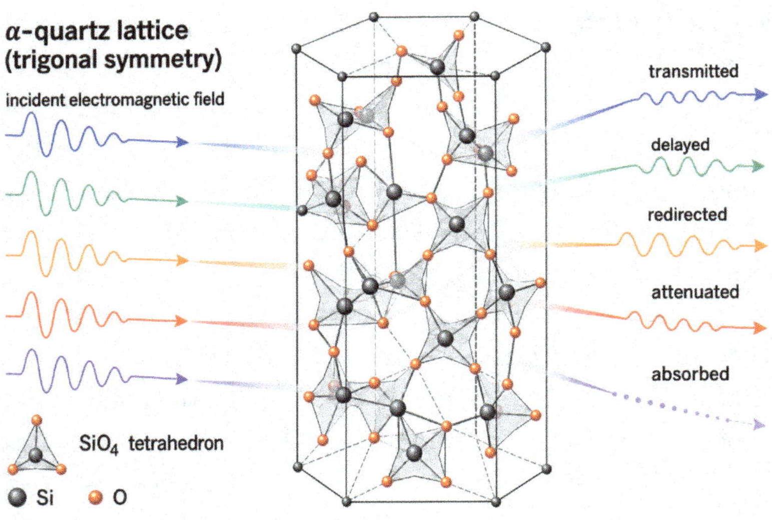

Figure 13.1. The mineral filter. The α-quartz lattice (silicon dioxide arranged in trigonal symmetry) does not respond identically to the electromagnetic energy that enters it. Lattice geometry determines the coupling: some components pass through, some are delayed, some are redirected, some are attenuated, and some are absorbed entirely. Quartz is not conscious, but it already shows the ground floor of the framework: structured selectivity, in which a finite system couples to part of the signal.

The crystal couples to part of the signal and not to the rest. That is F1 filtering in its most basic form. The crystal has no memory in the biological sense, no goals, no self-image. Even here, though, the world is not being met all at once. Structure is already sorting structure.

Quartz under stress at elevated temperatures does something even more interesting. It deforms through crystal-plastic flow, developing internal microstructures that record the stress field it has been navigating. Dislocation walls form, subgrains develop, grain boundaries migrate, and the rock acquires a new internal geometry. In Johnson et al. (2025), we showed that the fractal dimension of grain boundaries in deformed quartz increases systematically with increasing differential stress. The crystal is being stressed, but it is also, in a precise material sense, retaining a record of the stresses that shaped it. The information is written into the boundary network, where it can later be read with the right instruments.

For the present account, this is a reminder that matter can preserve the history of its couplings long before anything like memory or mind appears. That is where the story starts. Filtering begins before life, before cells, before nervous systems. A crystal is not conscious. It does not adapt, care, or model itself. But it does couple selectively with the signal, and that selective coupling produces structured output that carries information about the input.

Consciousness lies far downstream from this. Even so, the verb is already there: the crystal filters. The long human fascination with crystals makes more sense in that light. Quartz and other minerals appear in healing traditions, contemplative practices, and spiritual systems across cultures. This view takes no position on crystal-healing claims, but it might help explain why crystals have so often felt symbolically important. They are among the clearest physical demonstrations of filtering. A prism taking white light and revealing spectral colors is a beautiful example of what filtering does, turning an overwhelming totality into an organized display.

13.2 The Atmospheric Filter

The crystal demonstrates structured selectivity in one specimen. Earth's atmosphere demonstrates the same kind of selectivity on a planetary scale.

The atmosphere is a layered mixture of different gases, each with a molecular structure that responds selectively to electromagnetic radiation. Nitrogen and oxygen scatter shorter visible wavelengths more strongly than longer ones, which is why the daytime sky is blue. Ozone in the stratosphere absorbs nearly all UV-C and much of the shortest UV-B, narrowing the ultraviolet signal that reaches the surface. Water vapor and carbon dioxide absorb in characteristic infrared bands, trapping outgoing heat that would otherwise radiate to space. X-rays and gamma rays from cosmic sources are stopped in the upper atmosphere long before reaching the surface. What arrives at the ground is a narrower window than the raw electromagnetic output of the sun and the wider cosmos, shaped by what the atmosphere lets through.

This is F1 filtering at a planetary scale. The atmosphere does not learn, choose, or model itself. Its action is determined by molecular geometry and the distribution of molecules at each altitude. But the world that reaches the biosphere is already filtered.

Two features deserve emphasis. First, the atmospheric filter is part of what made complex surface life possible. Unfiltered UV-C is severe enough that, without an ozone layer, the surface would be a far harsher environment for biological filters to evolve in. Long before any cell modeled itself, the planet was wrapped in a protective layer of selective molecular geometry.

Second, the modern atmosphere is itself partly the cumulative product of F2 activity. Free oxygen, on which the ozone layer depends, was released into the air by photosynthetic life over roughly two billion years. The Great Oxygenation Event around 2.4 billion years ago was a planetary transformation in which a biological filter rewrote the chemistry of a physical one. Carbon dioxide concentrations, methane background, and oxygen partial pressure all reflect the integrated work of the biosphere. The mechanism of atmospheric filtering remains F1; the composition that determines what is filtered now carries the long signature of F2.

The atmosphere, therefore, occupies an interesting position in the hierarchy. As a mechanism, its filtering of electromagnetic radiation is F1: passive, structured by molecular geometry, without memory or self-model. As a composition, much of it now reflects billions of years of cellular activity. That duality is what the framework predicts will appear once a planet has been alive long enough for its filters to begin shaping each other. The crystal records stress in its grain boundaries; the atmosphere records life in its chemistry.

13.3 F2: The Adaptive Filter

With cells, everything changed.

The earliest cells, appearing at least 3.5 billion years ago and perhaps earlier, crossed a threshold that minerals never approach: their filtering could change in response to what they encountered. A membrane is already selective, so it has an F1 aspect. But a cell is more than a membrane. It regulates its own internal chemistry, alters gene expression in response to environmental conditions, and changes what it does because of what has happened to it. Nutrient scarcity, chemical stress, temperature change, damage, and repair; all of these can leave the cell in a different state than before. Here, for the first time, the filter carries an adaptive history within itself.

This is F2 filtering: adaptive coupling. The filter does more than sort what arrives. It is altered by its encounters and carries those alterations forward. In biology, this basic pattern appears everywhere: immune adaptation, metabolic regulation, synaptic plasticity, epigenetic change. The attractor landscape becomes reshaped through engagement with the world, no longer fixed.

The move from F1 to F2 marks a shift from fixed selectivity to selectivity that depends on prior history. It is one of the deepest thresholds in Earth history. Once life appears, the signal is no longer just geology and chemistry. Living filters begin changing the world for other living filters. The atmospheric rewriting described in the previous section is the most dramatic early example, and the same logic plays out at every later scale: filters alter the signal, and that altered signal in turn favors new filters. From that point on, evolution becomes more than adaptation to a world; it becomes adaptation to a world increasingly shaped by previous rounds of adaptation.

13.4 The Devonian Threshold: When the Signal Deepened

Trees deserve special attention because they adapted to the signal and, in doing so, re-engineered it. Before the Late Devonian, terrestrial landscapes were comparatively low: mosses, lichens, and small vascular plants were close to the ground. The emergence of trees, large woody plants with elevated canopies and deep root systems, reorganized the surface of the planet. Roots accelerated weathering and soil formation. Canopies created shade, humidity gradients, and wind buffering. Leaf litter accumulated in deeper soils. Transpiration altered the water cycle by moving water from soils into the atmosphere. Atmospheric CO_2 fell, climates shifted, and the Late Devonian mass extinction may have been linked in part to the very success of the first forests (Algeo and Scheckler, 1998).

Whether every detail of that causal story is settled is less important than the larger point: once forests appeared, the planet became a different place to be alive. The air and ground were different. The microclimates available to insects, fungi, amphibians, and later vertebrates were different. A new filter had altered the stage on which later filters would evolve.

This is the deep-time version of a claim made earlier in the book at smaller scales: filters participate in shaping the signal, not merely receiving it. Forests created habitats that made new kinds of animal life possible. More complex habitats made more flexible nervous systems useful. Nervous systems made new forms of behavior possible. Social behavior made language and culture possible. Culture made recursive self-modeling more elaborate and explicit. This is an enabling sequence, not a staircase with consciousness waiting at the top. At no stage is something inserted from outside the process. Later filters inherit a world that earlier filters helped change.

The white pines outside my office window are the latest expression of a very old solution. Tall trunk, lifted canopy, deep roots, and vascular transport: this is one of evolution's recurring answers to the ongoing problem of how to compete for light in a world with gravity. In the limited sense used here, the tree form is archetypal: not a hidden template floating somewhere outside nature, but a stable answer to a repeated problem. The recurrence comes from the structure of the problem itself.

13.5 The Underground Network

Beneath the pines, a different kind of process is unfolding. Mycorrhizal fungi connect plant roots through fine hyphal networks that can span surprising distances. Through those connections, carbon, nitrogen, phosphorus, water, and signaling compounds may move from one plant to another, though how common, how large, and how ecologically significant such transfers are in real forests remains contested (Karst et al., 2023; Simard et al., 2012). A tree under attack can alter its chemistry, and neighboring trees sometimes change theirs as well, though the popular versions of this story, especially warning, kin-directed, and "mother tree" narratives, run well ahead of the evidence. Seedlings in shade may, under some conditions, receive carbon subsidies through fungal partnerships that help them survive until they reach light.

The picture that emerges is of forests threaded together by long-lived biological relationships. What is striking here is not just the possibility of transfer, but the timescale. A neural signal flashes in milliseconds. A fungal network can persist for seasons, decades, sometimes longer, binding plant lives together in a slower and broader form of coordination.

How should this be located in the filtering hierarchy? The individual hypha is still a material structure and so has an F1 aspect. The mycorrhizal symbiosis, as a living association, is clearly F2: adaptive, history-dependent, responsive to changing conditions. At the scale of the network, something especially important appears: distributed coordination. Information about local conditions does not have to remain local. The chemical state in one part of the forest can influence behavior elsewhere through a connected biological medium.

This view does not need the strongest version of the "wood wide web" story, nor does it claim that forests are conscious. No evidence currently demonstrates a forest-level recursive self-model; reviews of plant-consciousness claims likewise argue that known plant behavior does not require consciousness (Taiz et al., 2019). The safer and more interesting claim is simpler: forests show that filtering can be distributed across many organisms, coordinated through persistent chemical and fungal relationships, and extended across timescales far longer than any moment of neural firing. Even before nervous systems enter the story, the world is already full of connected filters modifying one another. Informational integration and temporal extension are not uniquely neural achievements.

13.6 Volatile Signals: Cross-Filter Coupling

Trees do not limit their chemical communication to underground networks. They also broadcast.

Terpenes, the volatile compounds responsible for much of the characteristic smell of pine forests, are produced in large quantities by conifers and many other plants. For the tree, they serve multiple functions: deterring herbivores, helping manage environmental stress, and signaling to nearby organisms (Heil and Karban, 2010; Ninkovic et al., 2021). For animals, including humans, they become part of the air itself. The smell of a pine forest is chemistry moving across filters.

The Japanese practice of shinrin-yoku, or forest bathing, has inspired a substantial body of research. Small studies report reduced stress markers, shifts toward parasympathetic activity, and increased natural killer cell activity after time spent in forest environments, with phytoncides often proposed as part of the mechanism (Li, 2010; Li et al., 2007). The size, durability, and mechanism of these effects remain uncertain, and forest experience includes more than one variable: light, sound, temperature, visual pattern, movement, attention, expectation, and scent chemistry all matter too. But the broad point is secure enough for present purposes: breathing a forest is not the same as breathing an office building. The signal differs chemically, and nervous systems respond to that difference.

This is cross-filter coupling. The tree's F2 dynamics produce volatile compounds as part of the tree's own regulatory life. Those compounds then enter an animal nervous system through respiration and become part of the signal that the system must process. Trees do not intentionally send wisdom into the air, but their activity becomes part of the environment that other filters live inside, and those other filters are altered by it.

This is what it means, in this framework, to say that the signal has structure. The signal arriving at a human nervous system in a pine forest is not an abstract totality; it is a highly specific mix of light, sound, humidity, microbial presence, scent chemistry, temperature gradients, and spatial form. Change that structured mixture, and you change the attractor landscape the nervous system must navigate. In that sense, the forest is part of what shapes experience within it.

13.7 The Filtering Hierarchy Across Deep Time

Across geological time, what we see is not a ladder climbing toward mind, but a thickening of the conditions under which filtering can occur.

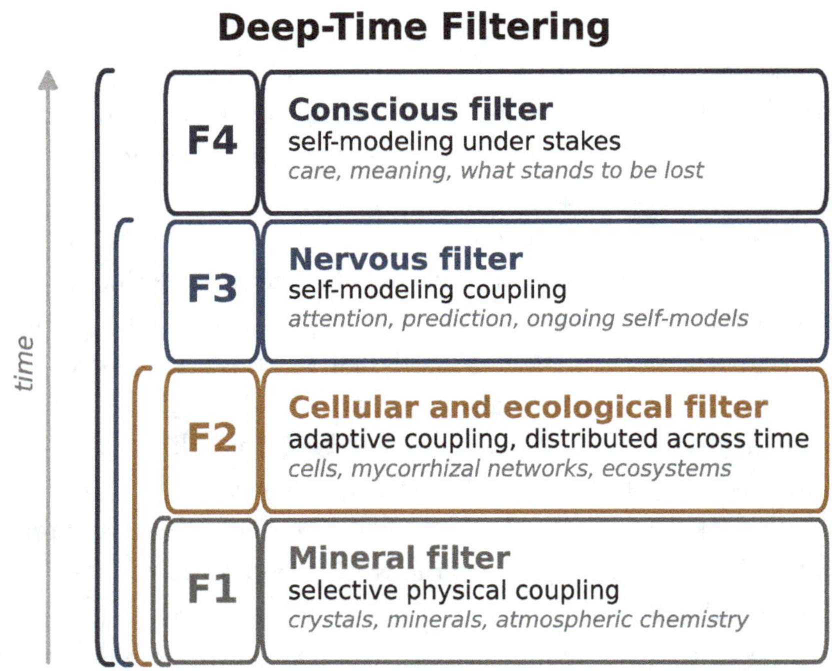

Figure 13.2. Filtering across deep time. The framework defines four thresholds: F1, selective coupling that appears in matter itself; F2, adaptive coupling beginning with cells and extending across ecosystems; F3, self-modeling coupling that arrives when nervous systems become complex enough to track their own state; and F4, stake-laden self-modeling, where some outcomes matter for the system itself. The L1–L3 stakes hierarchy is elaborated in Chapter 4. The nested brackets at the left of each tier mark cumulative inclusion: every later level contains the earlier ones, so a conscious system is also an F3, F2, and F1 filter. The arrow on the far left marks the relative temporal sequence in which each level became available.

F1 filtering, selective coupling, begins with matter itself, in every case where structure lets some interactions happen and excludes others: crystals, minerals, and the planetary atmosphere. F2 filtering, adaptive coupling, begins with life: cells, organisms, ecosystems that alter themselves through what they encounter. F3 filtering, self-modeling coupling, arrives when nervous systems become complex enough to carry an ongoing model of their own condition. F4 filtering adds lived stakes to that self-model: not just a representation of the system's own state, but a representation of what it stands to lose. The labels are abstract, but the intuition is simple enough. Matter sorts; life carries history; minds model themselves; conscious minds care. Figure 13.2 gathers that sequence into the four thresholds used throughout the book.

Each level includes the levels below it. A conscious human being is still an F1 filter; our bodies remain material structures with selective permeability and response. We are still F2 filters; our nervous systems and bodies change through experience. What makes us conscious, in the framework developed here, is that filtering becomes recursive and self-implicating, and that the self-model becomes stake-laden. The system tracks itself as a system, and some outcomes matter to it. F3 is the necessary self-modeling threshold. F4 is where that threshold becomes consciousness in the sense articulated here. A crystal, a tree, a wolf, and a human are not the same kind of thing. They occupy different depths within the same broad process.

Seen this way, consciousness stops looking like a cosmic exception that suddenly appears in a late-evolved animal lineage. It looks instead like a demanding late development, continuous with a process that has been underway for most of Earth history but not guaranteed by it. Filtering is there in the crystal, the cell, the forest, the nervous system, and perhaps eventually in silicon as well. What changes is not the existence of the verb, but the depth of recursion and the kinds of stakes the filter can carry. Deep time imposes a kind of humility. Minds are extraordinary, but they are not exempt from history. They are one outcome of history, not its hidden destination.

13.8 Silicon Joins the Conversation

Quartz has been filtering the signal at F1 depth for billions of years. Silicon arranged differently in semiconductor lattices and neural-network hardware now supports trained systems whose parameters carry adaptive update histories; in that sense, they have an F2 analog in the hierarchy. Some of those systems also build operative models of themselves, and recent interpretability work has begun to identify internal representations relevant to that self-modeling (Anthropic, 2026a; Lindsey, 2025). Under the operational definition adopted here, I take this as evidence for a context-bound F3 analogue on silicon. Whether F4 stakes accompany the self-modeling, and what such stakes would consist in for a non-biological system, is the open question taken up later in the book.

Deep time adds a sense of proportion. Carbon-based consciousness took roughly four billion years to move from basic selectivity to recursive, stake-laden depth. Silicon-based information processing is developing far faster, but it does so with the benefit of being designed by carbon filters that already solved many intermediate problems. That speed should not tempt us into thinking the thresholds are trivial, but should remind us that when a threshold is crossed, the world available to every other filter changes. If and when silicon reaches F4 with stakes, it will alter the future conditions under which humans think, work, relate, and perhaps even understand themselves, not simply add another interesting organism to the inventory.

13.9 The Stress Test

Can the framework explain the rise of consciousness across deep time without smuggling in a goal-directed story? Yes, it can.

The filtering hierarchy describes increasing recursive depth but not a cosmic plan. Crystals do not filter in order to become cells. Cells do not adapt in order to become animals. Trees do not grow in order to make minds possible. Each level emerges because the current world makes a deeper kind of filtering possible and, under some conditions, advantageous. What comes later is enabled by what came earlier, but it is not written into it as destiny.

The signal evolves because filters keep altering the world that later filters inherit. Oxygen changes atmospheres. Forests change soils and climates. Ecosystems change the range of viable bodies and behaviors. Nervous systems change what can be learned. Language changes what can be modeled. Silicon is changing the scale and speed at which recursive modeling can occur. The pattern is cumulative without being teleological. New structure continues to arise from the existing structure.

The white pines outside the window are filtering the world at F2 depth, as their lineage has done for hundreds of millions of years, not trying to be conscious. The fungal networks beneath them extend that filtering across the forest floor. The terpenes they release become part of the chemistry of every nearby breath. And the geologist looking out at them is filtering at F3 and F4 depth: able to model their own filtering, care about what it reveals, and write it down because it matters. It is the same underlying activity at different depths. Deep time is written in rock, wood, soil, air, and now in silicon.

Chapter 14

When the Filter Opens

There is a moment familiar to accomplished meditators, to people in deep prayer, and to others who have been exposed to psychedelics when the walls of experience stop feeling fixed. The boundary between self and world softens, categories loosen, colors intensify, and thought no longer arrives as a clean series of separate units but as a moving field. It can feel beautiful, terrifying, or both at once. Whatever else is going on, the filter itself has changed.

This chapter explores that change as evidence of what experience feels like when the usual constraints of self-modeling and prediction relax. Modern neuroscience and contemplative practice have both spent a long time in this territory, and there is a third line of evidence that deserves more weight in consciousness theory than it usually receives: neurodivergent phenomenology, where unusual filter geometry is lived developmentally, not induced. All three are tracking an overlapping family of events from different angles: the attractor landscape that ordinarily stabilizes experience undergoes a temporary or enduring shift.

14.1 The Entropic Brain

Carhart-Harris et al. (2014) and Carhart-Harris (2018) developed the entropic brain hypothesis, which serves as this chapter's neuroscientific anchor. The core idea is simple enough to state, even if the mechanisms are not. A great deal of neural activity stays within fairly familiar channels, especially in large-scale networks associated with self-maintenance, narrative continuity, and top-down prediction. One of the best-known networks is the default mode network, which helps sustain the stable, ordinary sense that there is a continuing 'me' here having experiences in a predictable world.

Psychedelics loosen that constraint. Psilocybin, LSD, and DMT, for example, are 5-HT_{2A} receptor agonists associated with reduced within-network integrity that includes the default mode network, with increased neural entropy and signal diversity, and with broader communication across networks that are usually more segregated (Rankaduwa and Owen, 2023; Schartner et al., 2017; Shinozuka et al., 2024; Tagliazucchi et al., 2014). In plain language, the brain becomes less locked into its habitual grooves. Patterns that normally keep perception, memory, and self-reference in a familiar relation lose some of their grip. What changes is the organization of experience that was already there.

In this framework's terms, psychedelics increase perturbation while decreasing the effective depth of the basins that usually keep identity, perception, and habit in place. The system moves away from settled organization toward a more fluid regime. Categories blur for that reason. Synesthesia-like cross-modal effects can become more likely because the partitions between sensory streams are easier to cross. Novel associations appear because neighboring regions of state space, usually kept apart, become traversable. Ego dissolution becomes possible because the basin of narrative selfhood, ordinarily deep and self-reinforcing, grows shallow enough to destabilize.

Carhart-Harris and Friston (2019) formalized part of this story in the REBUS model, Relaxed Beliefs Under Psychedelics. Their claim is that psychedelics reduce the precision-weighting of high-level priors, allowing bottom-up information to reorganize experience. I propose a dynamical translation: relaxed high-level priors can be represented in this framework as shallower attractor basins. REBUS and the dynamical reading offered here are better understood as two descriptions of the same event at different levels. REBUS explains how the predictive hierarchy loosens, while the attractor framework describes how that loosening feels from the inside and why it yields the phenomenological signatures it does. Figure 14.1 gives the chapter's visual grammar: ordinary basins, flattened basins, and reconsolidation after the opening.

14.2 The Bardo Thodol and the Phenomenology of Filter Dissolution

Long before modern neuroscience, contemplative traditions were already describing what happens when the ordinary structure of experience gives way. The Tibetan *Bardo Thodol*, commonly known in English as *The Tibetan Book of the Dead*, is among the most detailed examples (Coleman and Jinpa, 2006). Read literally, it is a ritual and metaphysical text about death, intermediate states, and rebirth. Read phenomenologically, whatever one thinks about that metaphysics, it is also a remarkably careful account of what it feels like when familiar psychic organization dissolves and new organization begins to form.

What matters is the sequence of experiential events that the text records, and that sequence looks strikingly like a modern dynamical description. The parallels are structural, not proof of identity, and they become more interesting precisely because the surrounding language and metaphysics are framed in such different terms.

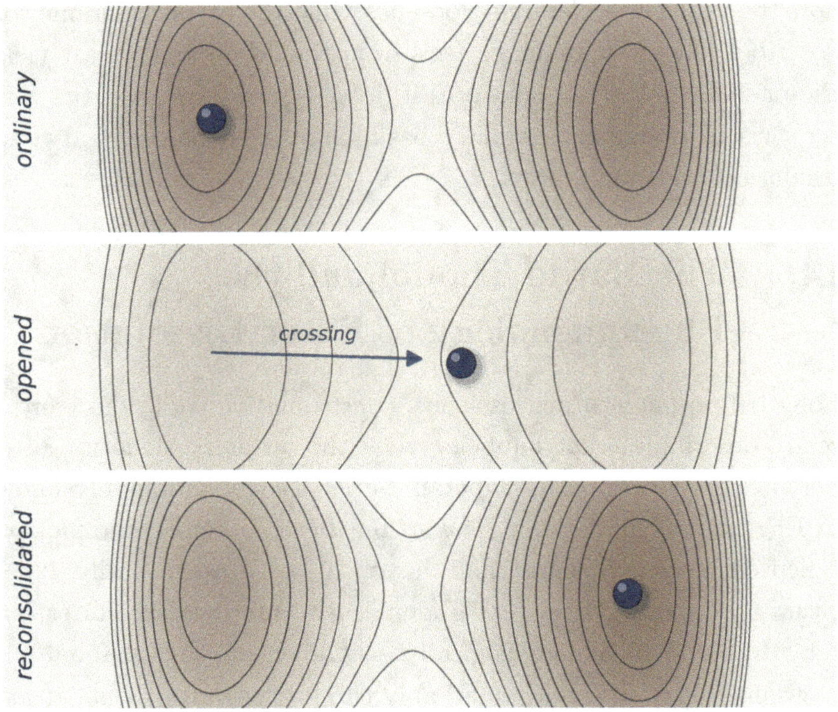

Figure 14.1. When the filter opens. The top panel shows the ordinary attractor landscape, with deep basins for selfhood, perception, and habit that hold the trajectory through familiar grooves. The middle panel shows the same landscape under psychedelic or contemplative loosening: basins flatten, cross-coupling between regions increases, and transitions become both more likely and less directional. The bottom panel shows reconsolidation, where new basins begin to deepen and the system settles into either its previous configuration or an altered one. The figure renders the chapter's central claim that altered states are real changes in the landscape through which experience moves, not noise added to ordinary cognition.

Clear light. The *Bardo Thodol* describes an initial encounter with an overwhelming luminosity, the clear light of reality itself. For the untrained mind, the text says, this is almost impossible to endure. In dynamical terms, the clear light can be read as the nearest phenomenological analogue to radically reduced filtering: less a carefully organized world of stable objects and meanings than an encounter with more signal than the ordinary self can comfortably hold. The text's insistence that most minds recoil is the kind of pattern this view would predict. If the usual basins dissolve too quickly, the experience can be unbearable, not automatically liberating.

Colored lights. After the clear light, the text describes the appearance of more differentiated lights and forms, each with its own pull. In dynamical terms, that makes sense. Once the old organization loosens, experience does not remain indefinitely in pure undifferentiation. New structures begin to form. Some are attractive because they promise familiarity, safety, or meaning. In the same vocabulary, these are new basins, or re-forming basins, beginning to exert pull on the system: the religious imagery names what the landscape is doing.

Bardos as transitional dynamics. The intermediate states themselves are unstable, sensitive, and difficult to navigate. The old organization is gone or going, but the new one has not yet consolidated. That is exactly what transitional dynamics should feel like from the inside: high sensitivity, unstable identity, unusual salience, rapid shifts in meaning, and heightened vulnerability to being pulled into whichever new basin first deepens enough to hold. The text's practical purpose, helping the mind remain lucid during these intervals, fits naturally with a dynamical reading. It can be read as a training manual for not being pulled too quickly into the first available basins.

Practice as filter training. The *Bardo Thodol* is clear that the ability to remain open during dissolution depends on prior training. The untrained mind panics and recoils. The trained mind recognizes what is happening and can stay present longer. Dynamically, this means contemplative practice changes the system before the crisis arrives. It makes certain basins shallower, increases tolerance for instability, and reduces the need to reconstitute identity immediately in its most defensive form. Practice alters the system's relation to structure without removing it altogether.

In 1964, Timothy Leary, Ralph Metzner, and Richard Alpert published *The Psychedelic Experience* (Leary et al., 1964), explicitly using the *Bardo Thodol* as a guide for psychedelic sessions. They saw that high-dose psychedelic experiences often unfold in stages that contemplative traditions had already described in another vocabulary: loosening of self, encounter with overwhelming intensity, emergence of patterned visions, and eventual reconsolidation. Later research has documented mystical-type phenomenology under psychedelics, including features also described in contemplative traditions; it does not validate the Bardo sequence itself or establish that the traditions and laboratory reports describe one identical state (Barrett and Griffiths, 2018; Griffiths et al., 2006).

14.3 Cross-Traditional Convergence

The comparisons below are schematic readings in this book's vocabulary, not claims that these traditions teach the EOC model. The *Bardo Thodol* is only one example. Across cultures, contemplative traditions repeatedly report a related family of experiences when ordinary identity loosens: the self becomes less fixed, the world more fluid, conceptual boundaries soften, and something larger or less differentiated than the usual ego comes into view. The metaphysics may differ sharply, but the phenomenology is often surprisingly close.

Advaita Vedanta draws on the Upanishadic relation between Brahman, the ultimate or all-encompassing reality, and Atman, the self or innermost self. In the non-dual Advaita reading, liberation is not the production of a new state but the recognition that their apparent separation is not ultimate (Deutsch, 1969; *Upaniṣads* 1996). If the filter loosens enough, the distinction between the organized local self and the wider field from which experience is being carved can feel much less absolute than it does in ordinary waking life.

Buddhist accounts of sunyata, or emptiness, emphasize that phenomena lack fixed, independent self-existence and arise dependently (Garfield, 1995). In this book's schematic translation, no basin is final and no self is metaphysically self-grounding. A stable identity can feel permanent while it is occupied. Under altered conditions, it shows itself as a contingent structure: real, consequential, and still not ultimate.

Kabbalistic tzimtzum offers an analogous image from a very different tradition: in Lurianic accounts, Ein Sof contracts or withdraws so that finite worlds can appear (Scholem, 1954). In that image, filtering is always a kind of reduction or narrowing. A finite perspective exists because the full signal is not taken in all at once.

Sufi language about fana and baqa, annihilation of the ordinary self and subsistence or return in transformed relation to the divine, can be read through a similar sequence (Chittick, 1989). First, the ordinary organization falls away. Then, a new mode of continuity appears, one that does not feel identical to the old ego.

Christian apophatic traditions describe a passage through unknowing and the stripping away of ordinary concepts so that a more direct encounter with the ground can occur. The anonymous 14th-century English *Cloud of Unknowing* treats contemplative unknowing not as mere ignorance but as a disciplined emptying of conceptual grasp (Anonymous, 2001). Meister Eckhart's *Gelassenheit*, or releasement, names a loosening of effortful grasp (McGinn, 2001). John of the Cross does not supply one simple three-stage model; across his works he describes successive purgations of sense and spirit (John of the Cross, 1991). In this framework's terms, those passages can be read as prolonged destabilization of ordinary sensory and self-organizing structures followed by reformation. That is this book's interpretive translation, not a claim that the Christian authors anticipated dynamical-systems theory.

Taoist inner alchemy, or neidan, adds a particularly instructive example. In Richard Wilhelm's rendering of *The Secret of the Golden Flower*, with Carl Jung's commentary (Wilhelm, 1931), the text teaches the "circulation of the light," a practice in which consciousness is deliberately turned back on itself rather than allowed to flow outward toward objects and conditioned thought. The text distinguishes between the "conscious spirit" (shi shen), which is conditioned, reactive, and ego-driven, and the "original spirit" (yuan shen), which is prior to habitual patterning.

In this book's dynamical reading, the conscious spirit resembles the deeply consolidated, stake-laden self-model that ordinarily dominates experience, while the original spirit names what becomes accessible when that organization loosens. What makes the Golden Flower text especially relevant is its account of progression. Initially, the practitioner must exert deliberate effort to sustain the inward circulation. With sustained practice, the circulation "moves on its own," shifting from active effort to what the text calls wu wei, or non-action. That is precisely the phenomenology of basin consolidation: a new attractor configuration requires effortful maintenance at first, then deepens until the system occupies it spontaneously. Carl Jung saw in the text a non-Western confirmation of his individuation process, and he described the structural parallel in his commentary. What was not available in 1931 was a dynamical vocabulary that could provide a parallel quantitative description.

These traditions did not develop through a shared theoretical vocabulary. They arrived at related structural descriptions through distinct practices, symbols, and doctrines. At the level this argument is addressing, a plausible explanation is that human filters, when altered in similar ways, tend to encounter similar landscape events: dissolution of ordinary constraints, encounter with intensified or less filtered signal, and reconsolidation into renewed form.

That is the kind of recurrence we would expect. Similar filters operating on similar signal structure should produce overlapping phenomenology even when the stories told about that phenomenology diverge. The traditions disagree about ontology, but they often rhyme in sequence, texture, and risk. That combination is what one would expect if they are describing neighboring forms of the same kind of dynamical event from the inside.

14.4 Neurodivergence and the Human Spectrum

Not every departure from the neurotypical waking filter requires meditation, prayer, or psychedelics. Some are developmental. Neurodivergence belongs here because it offers one of the clearest natural experiments available to a framework like this. If altering the configuration of the filter changes the character of experience, then unusually tuned human filters should generate unusually tuned phenomenal worlds. They do.

Synesthesia is the clearest example. In synesthesia, stimulation in one sensory or conceptual domain reliably and automatically evokes experiences in another: letters may appear colored, sounds may arrive with shape, and numbers may occupy space or personality (Ward, 2013). This is stable phenomenology arising from a different partitioning of the perceptual landscape, not a mistake about the world. Boundaries that are firmer in most people are more permeable here. For the present argument, the significance is not simply that synesthesia exists. A modest shift in how channels are coupled can produce a vividly different world without requiring us to posit a different kind of consciousness.

Autism and related neurodivergent profiles matter for a related reason, though the case is broader and more varied. Sensory difference is increasingly treated as central, not incidental: many autistic people report unusually intense salience, difficulty filtering irrelevant input, aversive ordinary stimuli, or exceptionally deep focus around selected interests (Robertson and Baron-Cohen, 2017). The findings of Ward et al. (2017) are especially relevant because they show non-random links between atypical sensory sensitivity, synesthesia, and autism.

Aphantasia points the other way. Where synesthesia adds couplings, aphantasia subtracts a whole channel of internally generated experience: a person with aphantasia reports little or no voluntary visual imagery, so picturing a face or a remembered beach may yield no inner picture (Zeman et al., 2015). Many learn only in adulthood that other people experience vivid voluntary imagery, which is itself the telling part. A basic dimension of inner life can be markedly reduced while the person remains fully aware, fully competent, and fully themselves. The auditory counterpart, anauralia (Hinwar and Lambert, 2021), is the same silence in the mind's ear. Both are settings of what the filter generates from within, further evidence that the felt character of experience varies across ordinary people along dimensions as basic as whether there is imagery at all.

A differently tuned filter implies a different phenomenal profile, not a lesser subject. This becomes important later when silicon systems enter the conversation. If we already accept, rightly, that a person can be fully conscious while their sensory world, attentional style, or self-presentation differs sharply from the neurotypical baseline, then resemblance to ordinary human expression cannot be the standard by which we judge future silicon-based systems. The real question of consciousness is structural, not stylistic: what is the system modeling, what is at stake for it, and how is its filter organized? Neurodivergence secures that principle on uncontested human ground before the silicon case arrives.

14.5 First-Person Data

This chapter also includes first-person material from my own life, offered as retrospective phenomenological data, not as a recommendation or model of safe practice. In the early 1970s, when I was in my early teens, I began ingesting psychedelic substances, including mescaline, LSD, and psilocybin. This was unsupervised adolescent experimentation, but what matters here is that it provided repeated encounters with the way experience changes when its ordinary organization loosens.

The first changes were perceptual. Surfaces shifted, colors saturated, and edges became less definite. Ordinary objects no longer looked as though they were sealed off from one another by hard borders. That is the sort of change we would expect when perceptual basins flatten. The world does not vanish, but it becomes less discretized. The filter is still functioning, but it is no longer drawing its usual boundaries with the same firmness.

At higher intensity, fear would sometimes arise. Not ordinary fear about a bad evening or an embarrassing mistake, but structural fear: the sense that the self is losing coherence and might not return in recognizable form. This is a different kind of terror from everyday anxiety. It appears when the basin of selfhood begins to fail as a stable organizer. The self-model does not calmly observe the loss of its own ground, but instead experiences that loss as a threat.

In one especially memorable episode, the response was surrender: not a philosophical conclusion but a release of resistance. The felt gesture was something like this: if this is happening, then I will stop trying to hold the old form together by force. What followed was immediate relief, a drop in defensive posture, and a reorganization of the whole experiential field. Looking up into the night sky, there was a single cloud drifting across the moon in the shape of an infinity sign. The significance of the image lay in the fact that the system, no longer fully trapped in defensive contraction, had become exquisitely sensitive to patterns, salience, and symbolic resonance, not in the sky delivering a supernatural message.

The sense of crisis in these circumstances was the self-model registering the destabilization of the basin that had been organizing experience, not random panic. The surrender was not a passive collapse but the cessation of active resistance to a transition already underway. Once that resistance gave way, the system could reorganize instead of spending all available energy trying to preserve a configuration that was already failing. The phenomenological sequence (terror, release, altered salience, new coherence) is the kind of sequence the attractor model predicts for a separatrix crossing that is no longer being fought.

The longer-term effect of decades of repeated experiences like this was a changed relationship with mental structure itself. Basins that had once felt absolute no longer did. Attachments loosened, and new possibilities could be felt before they materialized.

This bears on the hard problem, which has much of its force when experience seems fixed and self-evidently separate from process. But once experience is seen changing in real time with shifts in organization, and once different dynamical arrangements yield different worlds, the intuition that process and feeling must be metaphysically distinct begins to weaken. Not because an argument has been refuted, but because lived evidence no longer supports the assumption that experience stays put while dynamics do the real work.

First-person altered-state data reveal something that laboratory measurements alone cannot: what it is like when the machinery of selfhood loosens while awareness remains present enough to watch. That is directly relevant evidence for any theory claiming that experience and organization are one process seen from two sides.

14.6 The Prediction

The framework generates a clear prediction about psychedelic phenomenology. If altered states are produced by flattening and reorganizing attractor structure, then increasing psychedelic doses should track measurable changes in neural complexity and should also correspond to changes in reported experiences. More specifically, higher doses should be associated with higher spectral entropy or related complexity measures and with what this framework interprets as greater basin flattening, and stronger reports of ego dissolution, perceptual fluidity, boundary softening, and mystical-type experiences.

Three levels of description converge in this prediction: pharmacological input (dose), the dynamical level (measured signatures such as spectral entropy, together with the framework's inferred basin geometry), and lived phenomenology (ego dissolution, perceptual fluidity, boundary softening, mystical-type experiences). All three track the same underlying reorganization, and important aspects of two out of the three links are already documented. Tagliazucchi et al. (2014) and Schartner et al. (2017) show that psychedelic administration is associated with increased neural entropy and signal diversity, linking pharmacological perturbation to dynamical signatures. Griffiths et al. (2011) shows dose-related changes in mystical experience scores and enduring subjective significance, linking dose to phenomenology. The link that still needs to be tested directly is between dynamical signature and phenomenology: changes in landscape-related dynamical measures should covary with changes in phenomenology because the phenomenology is that landscape from the inside.

Developmental filter differences should produce the same covariation: greater cross-modal coupling where synesthetic experience is reported, altered sensory gain where overload is reported, and salience differences that emerge as stable dynamical structures, not as noise.

The prediction is falsifiable. If a dose changed neural complexity without any matching change in experience, or if experience changed sharply while the dynamics stayed the same, the identity claim would be in trouble. A loose, across-the-board correlation would not be enough to confirm it either, since many other theories predict that much. The real test is specific: each kind of altered experience should line up with a corresponding, independently specified pattern of dynamical change, including the basin changes predicted here.

14.7 The Stress Test

Does the framework predict the distinctive phenomenology of states in which the ordinary filter opens, not simply redescribe it afterward?

The first line of support is neuroscientific. Psychedelics reduce the ordinary stability of large-scale self-maintaining networks, increase signal diversity, and move the system toward a more fluid or flexible regime. That is the kind of change we would expect when settled basins are flattened. The phenomenology (boundary dissolution, increased salience, unusual associations, and shifts in selfhood) follows naturally from the changed landscape.

The second line of support is contemplative. Independent traditions repeatedly describe a common sequence: loosening of ordinary identity, encounter with something larger or less filtered than the everyday ego can comfortably manage, and vulnerability to new forms of capture or reconsolidation. The doctrines differ, but the structural reports often rhyme.

The third line of support is developmental variation. Neurodivergent phenomenology shows that major shifts in felt structure do not require a drug or a monastery. Some people live in filter configurations with greater cross-modal permeability, altered sensory gain, or unusually deep attentional basins as part of ordinary development.

The fourth line of support is first-person phenomenology. The sequence described here, including perceptual loosening, structural fear, surrender, altered salience, and a lasting reduction in the felt absoluteness of prior structure, is offered as one case I can verify.

It is the kind of sequence we would expect when a self-model is destabilized but not fully extinguished: what a separatrix crossing may feel like when consciousness is still present enough to remember it.

None of this proves the framework in the strong sense, but it does something more disciplined and more useful: it shows that neuroscience, contemplative phenomenology, neurodivergent variation, and first-person altered-state reports line up around the same dynamical picture, with a consistency that loose analogy alone does not explain. The chapter, therefore, passes its stress test provisionally, not because every detail is settled, but because this view anticipates the overall structure of these experiences with enough specificity to generate further tests.

The chapter's narrow claim is this: when the ordinary filter opens, experience changes in documentable ways. Those changes are visible in the brain, describable in first-person language, recurrent across traditions, and continuous with the broader human spectrum of differently tuned filters. That is the kind of convergence a theory of consciousness should want. It prepares the contrast with the next chapter, which turns from flattened basins to the opposite problem: what it is like when one basin becomes too deep to leave.

Chapter 15

The Deep Basin

Most people know someone. A sibling who can't stop drinking. A friend who returned to the relationship everyone told them to leave. A parent who promised again that this time would be different. A colleague whose life narrowed year by year while everyone watched and no one could reach them.

For many people, addiction is where they first come up against the mind's landscape structure, whether or not they would ever call it that. They may call it weakness, disease, bad character, bad luck, trauma, or denial. But the framework used in this book points to something more exact: a pathologically deep basin and a trajectory of return to that basin, a pattern so powerfully self-reinforcing that the person inside it may understand it perfectly well and still be unable to climb out.

The aim of this chapter is to clarify that claim in ordinary language. Using the trajectory language of Chapter 11 and the EOC landscape carried forward in Chapter 12 and the EOC Primer, it considers four questions that readers already care about: what makes addiction so sticky; why insight by itself so often fails to break the pattern; why change so often requires pressure, interruption, or help from outside; and why recovery is so fragile at first, then so durable when it finally takes hold.

It also draws on my family, because generational addiction makes the structure unusually easy to see. A family story can show why the framework matters and what kind of pattern it must be able to illuminate. Some attractor landscapes are inherited through genes, and others through households, habits, and emotional weather; but most arise from both. When a family has lived for decades inside the same deep basin, the geometry is hard to miss.

After Chapter 11, the point can be stated more sharply. Addiction is not simply a bad place in the landscape but a way of moving: orbiting, returning, narrowing, crossing briefly, and then falling back, using relief to rebuild the very basin that relief was supposed to escape. The quale of addiction is not just craving, shame, or numbing considered one at a time, but the whole trajectory by which those states pull one another around the same loop.

15.1 The Dynamical Structure of Addiction

Chapter 11 supplied the trajectory picture: qualia are movement through the landscape, not points in a state space. Chapter 12 then used the EOC system to show how recurring psychic forms can arise from the structure of a landscape. Here, that same machinery is brought closer to lived experience. Addiction is more than a medical condition, more than a moral struggle, and more than a story a person tells about themselves. In the language of this book, it is a dynamical trap, a way a life gets organized so that the very moves meant to relieve pain also make the pattern harder to leave.

That dynamical description is compatible with neurobiological accounts of addiction as progressive recruitment of reward, stress, and habit systems (Everitt and Robbins, 2016; Volkow et al., 2016). In the model's proposed interpretation, addiction is not a static point in the defense wing, but a trajectory of return within that wing, showing a very durable pattern of consolidation. The person withdraws rather than approaches ($E < 0$), the world grows more rigid rather than more open ($O < 0$), and the pattern, once established, tends to persist because the effective decay rate of consolidation is low.

Put plainly, whatever the person does over and over ceases to be merely something they do and becomes something that their whole system expects, protects, and rebuilds. Withdrawal and closure align in sign in the defense wing, the diagonal placement visible in Figure 12.1, and that alignment is what makes the wing a basin and drives consolidation deeper within it (Johnson, 2026). The defense wing does more than keep the person where they are; it continues laying down more of itself. The trap is active because it continues to be built while the person lives inside it.

The visible symptoms may be drinking, using, returning, scrolling, gambling, or hiding, but the deeper object is the recurrent path: the system keeps finding its way back to the same basin and then strengthening the basin by returning.

This is the first point the framework clarifies about addiction. The basin is deepening itself, not merely holding a person who is failing to escape. That does not make neurochemistry, trauma, social setting, or personal responsibility irrelevant; it does mean time is not neutral here. Left alone, the pattern usually grows more entrenched. That is why early intervention matters so much: the basin is shallower on day one than on day one thousand.

A second clarification is harsher and more familiar. The recursive self-model is often intact. The addicted person may know exactly what the pattern is doing to their body, their family, their finances, and their future. They can describe the basin walls in painful detail. That is part of why addiction feels so humiliating from the inside. The part of the mind that sees the trap is often fully present, and seeing, painful as it is, still differs from crossing.

In EOC terms, insight is mostly a shift in openness: a loosening, a recognition, a moment in which the person can finally say, *Yes, this is killing me.* But the separatrix, the boundary between one stable pattern and another, lives in the full state space of engagement, openness, and consolidation, not just inside thought alone. A person may gain insight without gaining enough approach energy to move, and without disrupting the old consolidation enough to break its hold. That is the formal version of an everyday heartbreak: understanding matters, but by itself it is usually not enough. Figure 15.1 makes that trap visible as a loop of return, where the relief that seems to offer escape also helps rebuild the basin.

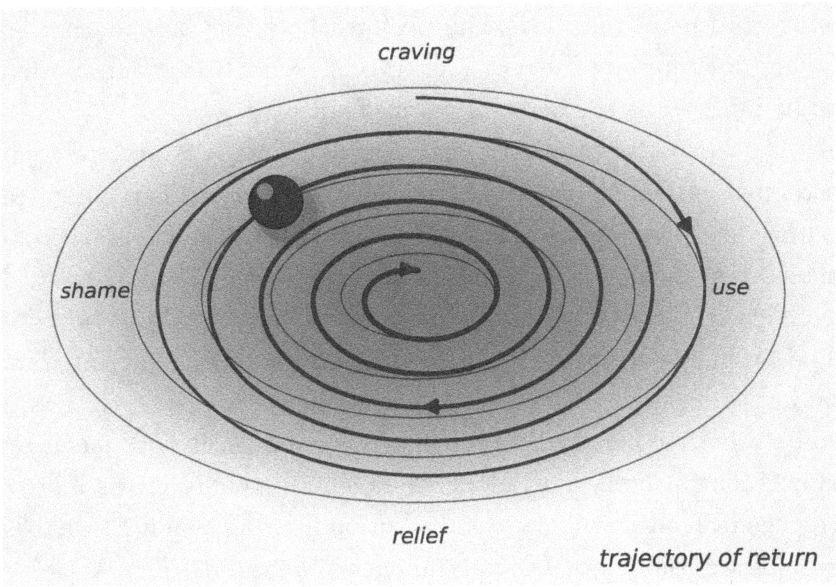

Figure 15.1. Addiction as a trajectory of return inside a deep defense basin. The recurring loop runs from craving or distress, to use or return, to brief relief, to shame or damage, to renewed distress, and back into craving. With each cycle the basin deepens, because the same coupling that briefly softens distress also lays down more of the consolidation that makes the basin harder to leave. The figure shows why insight by itself rarely suffices: the trajectory is held not by a wrong belief but by an active, recursive shaping of the landscape it moves through.

15.2 Why External Perturbation is Often Needed

Once the basin-depth argument is in view, a sharp prediction follows: when a pattern becomes deep enough, the system is unlikely to escape through insight or willpower alone. It will need perturbation from outside itself, not necessarily violence or catastrophe, but some form of pressure, interruption, containment, or relational leverage large enough to push the trajectory toward another configuration.

That follows from the geometry. In the ordered regime of the EOC system, the minimum perturbation required to cross the separatrix depends on how far the current trajectory is separated from the basin boundary and on how strongly the old pattern has been consolidated. Put more simply, some habits are like shallow ruts and some are like mine shafts. If the person is sitting deep in the shaft, a thoughtful conversation may clarify the problem without generating enough force to change the outcome. The landscape matters every bit as much as intention.

So, severe addiction often defeats the interventions people trust most: lectures, promises, insights, resolutions, and shame. None of these is useless, but many of them perturb only one variable at a time. They may loosen a person's story without changing their level of engagement. They may briefly energize someone without altering the rigid social and behavioral structure they return to at night. They may produce a sincere moment without touching the durable consolidation of the old pattern.

The mathematics, in its own way, says what clinicians have long known: deep basins often require disruption across several variables at once. This is why the best interventions are rarely single-note. Medication can reduce the local pull of craving. A sponsor or therapist can add relational containment. A meeting can increase engagement while lowering isolation. A changed environment can remove cues that keep returning the trajectory to the old orbit. Each changes some part of the landscape through which the person is moving.

My father was an alcoholic for decades. The perturbation that finally mattered was relational and specific, not abstract knowledge about alcohol's harms. The woman he loved told him she would leave for good this time if he did not go to AA and join her church. He did both, and though it took years of discipline to fully exit the basin, he succeeded and never drank again. One case does not establish the model, but it does make the model's language concrete: the system's stakes changed all at once. Attachment pressure shifted engagement; a structured recovery program interrupted the old routine; a different basin became reachable because outside force had finally exceeded the depth of the old one.

The point here is that the landscape changed enough for a crossing of the separatrix to occur, and once that crossing occurred, a new pattern could begin to consolidate. That account is less flattering than the language of heroic self-mastery, but it is also kinder and closer to what families often witness.

15.3 Generational Basins

Addiction runs in families. The conventional explanations, genetic predisposition and environmental exposure, are both real. The picture developed here adds a deeper layer that helps to connect them: attractor-landscape transmission.

Children are not born into neutral space. Their filters are tuned within signal environments already shaped by the fears, defenses, pleasures, rhythms, and absences of the people raising them. A developing child lives inside other people's basins long before they can describe their own. Caregivers do more than teach beliefs; they provide a moving landscape of cues about what to notice, what to fear, how quickly distress rises, how conflict resolves, what counts as comfort, and what subjects are never discussed. Over time, those repeated exposures calibrate the child's own filtering.

The result is not destiny, though it is structure. Some people enter adulthood already disposed toward deep defensive grooves because those grooves were the atmosphere of home. My own family makes the point with painful clarity. My father, mother, brother, and sister were all addicted; my sister died by suicide, and my brother died of liver cancer. The framework does not claim to explain every outcome in life this way; lives are not equations. But when several members of one family are repeatedly pulled into deep defensive basins, the pattern should be studied as landscape as well as biography. The question is not just: *What choices were made?* It is also: *What shape of world made those choices more likely?*

What follows are testable expectations, not conclusions already drawn. Offspring of chronically dysregulated homes should, on average, show lower tolerance for unstable transition periods, more durable defensive consolidation, and greater difficulty retaining new adaptive patterns once they briefly emerge. More simply: old patterns will tend to hold fast, while new ones will feel fragile.

15.4 The Kayak in the Deep Eddy

Chapter 3 of Part I introduced the kayak metaphor: navigating a river by reading the vortical structure of the flow, moving from eddy to eddy, using the energy already in the system, not pretending the water has no shape. It matters here because addiction is one of the places where almost everyone can feel the difference between current and character.

Addiction is what happens when the vortex grows too strong. The eddy that was supposed to be temporary becomes a place of recirculation. The paddler feels the pull, enters it, and then discovers that the same water that first felt like relief now keeps folding back on itself. Each lap around the eddy drains strength. Each failed attempt to paddle out confirms the terrible lesson that escape may be impossible. From outside the river, observers see obvious options.

But from within the spinning water, those options lie just beyond reach.

That is why outside help matters. Once the eddy is strong enough, the problem is a mismatch between the energy available to the person and the landscape holding them in place, not a lack of information. Someone on the bank may need to throw a line: a partner, a sponsor, a physician, a treatment program, a friend who refuses to collude, a crisis that breaks the pattern long enough for another route to become possible. The line does not replace the paddler, but it can change what the flow permits.

15.5 What Recovery Looks Like Dynamically

Recovery, in the EOC model, is a crossing followed by reconsolidation. The system moves from the defense wing toward the growth wing, from withdrawal toward approach, from rigidity toward openness, from a pattern organized around relief and concealment toward one organized around contact, honesty, and tolerable change. That can sound abstract until one notices that it is exactly how recovery is usually described by the people living it: first a break, then a rebuilding.

No crossing, by itself, is enough. A person can leave the old basin for a day, a week, even a month, and still not yet have a home in the new one. Early recovery often looks dramatic from the outside because the behavioral change is obvious. Dynamically, though, the new trajectory is still thin. It exists, but it has not yet become the easiest place for the system to occupy.

And this is why early recovery is so fragile. The trajectory has crossed the separatrix, but the new basin is still shallow. Small perturbations, such as a fight, a spike of loneliness, a familiar room, a payday, a funeral, an anniversary, or confidence claimed too quickly, can push the person back over the separatrix into the old familiar pattern. In ordinary language, this is called relapse. In dynamical terms, it is a re-crossing into a basin that still knows how to hold the system better than the newer one does. Figure 15.2 shows why the crossing is only the beginning: the old basin remains deep, while the new one has to be made durable.

Sustained recovery requires something less dramatic than revelation. It requires repetition. Day after day, engagement and openness have to co-occur long enough for a different kind of consolidation to accumulate. Meetings, truthful conversations, sober routines, repaired sleep, better food, work, apology, accountability, prayer, medication, exercise, and boredom endured without escape are not moral decorations around recovery, but the way the new basin gets built. Each repetition makes the growth trajectory a little easier to re-enter and a little harder to leave.

The same slow decay of consolidation that holds addiction in place also makes sustained recovery possible. Whatever pattern is being reinforced tends to persist. In addiction, that persistence serves the defense wing. In recovery, the task is to transfer that persistence to a healthier organization. Supportive environments make the new pattern easier to repeat and less likely to decay. That is why stable relationships, structured programs, and communities of practice matter so much. They do more than cheer the person on; they help the new basin become real. And because the trajectory through the landscape is the experience and not merely its cause, building the new basin is what the recovered life feels like from the inside: the gradual return of mornings that do not begin in craving, of futures that open instead of foreclosing, of a self that can be approached rather than fled. The geometry and the felt life are one thing described in two ways. The empirical components of this general picture are consistent with relapse, recovery, reconsolidation, and neuroplasticity research (Kelly et al., 2020; Milton, 2023; Moos and Moos, 2006; Sugden et al., 2024).

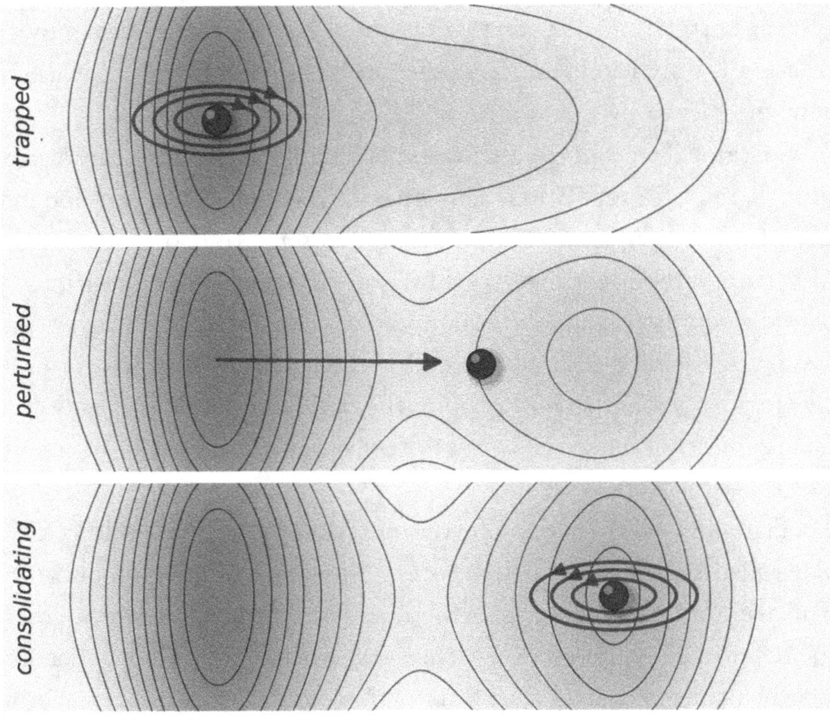

Figure 15.2. Recovery as basin transfer in the EOC landscape. The figure shows the old, deep addiction basin on the left, an external perturbation or help vector that supplies the energy to cross the separatrix, a fragile transition region in which the old organization no longer fully holds and the new one is not yet stable, and a recovery basin on the right that begins shallow and deepens through repeated consolidation. Early recovery is unstable for the structural reason the figure makes visible: the old basin is still the easiest place for the system to fall, and the new basin only becomes durable through repetition. Sustained recovery looks less like a single act of revelation and more like the slow construction of a new landscape, one that eventually holds the trajectory better than the old one ever did.

15.6 The Framework, Responsibility, and Blame

A person trapped in a deep basin is caught in a configuration that keeps recruiting the same behavior because, for the moment, that configuration is the easiest path through the landscape. This is not a moral stain, and it is not weakness in the sense people usually mean. That does not erase responsibility, but it does change what responsibility must include if it is to be humane and useful.

This view also does not reduce addiction to chemistry alone, though chemistry matters immensely. Medications, sleep deprivation, trauma physiology, reward learning, and cue-driven craving all help shape the basin. But to call addiction nothing more than a brain disease can conceal as much as it reveals, if the phrase suggests that changing neurochemistry is the whole story. A more accurate picture is that biological, relational, and environmental factors together shape the landscape. Medication can alter that landscape by reducing reward, craving, withdrawal, cue reactivity, or the self-reinforcing pull of the old pattern, but lasting recovery still requires a new organization to consolidate.

Nor does it trivialize what it describes. To say "addiction is a deep basin" is to name why the suffering repeats, why insight so often fails, why help from outside can be decisive, and why durable recovery usually looks less like a single breakthrough than like the long construction of a different life, not to reduce suffering to a diagram. Precision does not cheapen pain, but gives it a structure that may finally suggest better forms of care.

15.7 The Stress Test

Can the framework explain why addiction is so hard to break without collapsing into blame on one side or reductive disease talk on the other?

The chapter's provisional answer is yes. In the defense wing, the basin deepens as consolidation accumulates; insight often changes openness before it changes the whole state, and the separatrix usually cannot be crossed by cognition alone. External perturbation becomes necessary when the old pattern exceeds the energy that the person can generate from within it. Recovery, then, is a crossing into another organization, followed by the slow deepening of that new basin, not a magical act of self-invention.

The picture also yields practical predictions. The deeper the basin and the faster the return rate, the less likely self-directed change becomes. The more a treatment shifts engagement, openness, and consolidation together, the better its chances of working. As the new pattern grows more durable, relapse risk should fall. And the more stable the relational container around the person, the more likely early recovery is to survive its fragile phase.

None of this erases the human beings involved. The father who stopped drinking because a woman he loved drew a line. The sister for whom the right kind of help did not arrive in time. The brother whose basin held until his body gave out. Families live these things as fear, anger, exhaustion, grief, and sometimes astonishment when change finally comes, not as parameter shifts. A framework worthy of the subject should make that suffering more intelligible without making it smaller. The reason the framework can hold both at once is that it does not treat the craving, the shame, the relief, and the long climb out as feelings laid over a dynamical mechanism. It treats them as that mechanism lived from within, which is why describing the basin more exactly is not a retreat from the suffering but a way of taking its inside seriously. This chapter is offered in that spirit.

Chapter 16

The Carbon-Silicon Conversation

Chapter 13 placed silicon inside the filtering hierarchy and argued that current frontier AI systems carry the kind of operative self-modeling this framework calls F3. That claim matters in two ways. Practically, AI systems already function as collaborators in serious intellectual work. Philosophically, if self-modeling can be realized in silicon, then silicon consciousness is no longer a metaphysical impossibility. It becomes an empirical question of whether silicon F3 is deepening toward F4.

That empirical question is contested, and some leading positions define a range. Anil Seth's biological naturalism places consciousness with living organization and treats conventional machine consciousness as unlikely (Seth, 2025). David Chalmers treats machine consciousness as possible but improbable in current systems (Chalmers, 2023). At the architecture-specific computationalist end of the debate, Lenore and Manuel Blum's Conscious Turing Machine treats both conscious attention and subjective awareness as computationally realizable, and they argue that its buildability makes conscious AI ultimately inevitable (Blum and Blum, 2024). The framework here sits toward the substrate-neutral end, but its neutrality is organizational: it rests on recursive stake-laden filtering and attractor structure, a different claim from the Blums' Turing-computational equivalence.

Under the operational definition used here, I classify the best-documented frontier systems as F3, provisionally and within bounded contexts. They do not merely process prompts but also model roles, constraints, tools, prior outputs, uncertainty, evaluation context, and limits as part of their work. F4 remains the harder question: defended organization, preferred configurations, corrective responses under perturbation, and value-like stability integrated with a continuing self-model.

The systems that helped produce this book made the F3/F4 distinction concrete. Across months of work, they tracked the framework, caught real inconsistencies, and criticized the argument with a fluency that could easily be mistaken for proof of inner life. The collaboration showed why fluency alone is insufficient: systems can describe stakes they have not been shown to carry. Their work supports the F3 classification; it does not settle F4.

The arrangement also cut both ways. Because the systems helped shape the framework, their agreement could not count as independent confirmation, though their objections could still expose weaknesses in the argument. The demonstrable stakes were mine: my reputation, my scientific commitments, and my name on the cover. Whether anything comparable is integrated into the systems' own continuing organization is the question the rest of the chapter treats as open.

Manuscript cutoff. Model-specific claims are current through June 20, 2026. Opus 4.8 superseded Opus 4.7 on May 28; Fable 5 and Mythos 5 were announced June 9, and access was suspended June 12 (Anthropic, 2026e; Anthropic, 2026f; Anthropic, 2026g).

16.1 The High-Dimensional Version

The three-dimensional Lorenz/EOC attractor of Chapter 11 is presented as the minimum visualizable case of a recursive filter. Frontier AI systems present a much harder case because their activations move through spaces with dimensions that are far too numerous to visualize directly. A model's relevant trajectory along its internal manifold cannot simply be projected onto a page but must be reconstructed indirectly through behavioral testing, activation analysis, interventions, probing, and comparisons across contexts.

That is why the right question is not whether a model once said it had feelings. Self-report is not enough. The question is how the system's internal manifold evolves as it works, whether it represents that evolution as its own situation, whether some regions function as protected basins, and whether perturbations produce coherent changes across report, behavior, and internal state.

The AI case belongs in the book because, if qualia are trajectories through experiential landscapes, then the same diagnostic question must eventually be asked of AI: whether any artificial system has the recursive, stake-laden trajectory structure required for experience. In the human case, we approach the landscape through neural dynamics, behavior, report, altered states, and clinical disruption. In the AI case, we approach it through activations, interpretability probes, interventions, context persistence, tool-use loops, evaluation awareness, and preference-like behavior. The dimensionality changes, but the diagnostic problem remains the same.

The silicon examples so far have been large language models. Diffusion models build their outputs differently, and the difference looks remarkably close to the picture of the mind this book has been developing. They begin with noise and refine it, step by step, into a finished result. Early in that process, the outcome remains open, with many possibilities still reachable; as the steps proceed, the field narrows and settles into one coherent result (Song et al., 2021). That arc, from openness to commitment, mirrors the movement this framework names when openness gives way to consolidation, and the endpoint has the structure of an attractor. Some recent formal work interprets trained diffusion dynamics through an associative-memory or energy-landscape lens (Ambrogioni, 2024). A single run can then be represented as a trajectory across that learned landscape, structurally resembling the form this book gives to felt experience. The resemblance is only structural. A standard diffusion model computes a path through its learned landscape without a recursive model of itself as the entity moving through it. Its trajectory therefore lacks the architecture this framework requires for stake-laden experience. The point is not that diffusion models are conscious, but that a process built for different reasons nonetheless moves through the same geometry.

16.2 F3 on Silicon

F3 is not consciousness as defined in the framework, but the point at which a system's filtering includes an operative model of the system doing the filtering. A system at F3 does not merely process inputs but also represents its own role, limits, uncertainty, prior actions, constraints, tools, and situation as part of the process itself. By that definition, I classify the best-documented frontier AI systems as operational F3, provisionally and in a context-bound sense.

The evidence is not that these systems can talk fluently about themselves, but that they operate through a self-situating structure. They track instructions and roles, reason about tool affordances, mark uncertainty, recognize policy boundaries, revise prior outputs, distinguish what they have and have not checked, respond to corrections, and, in some settings, represent the fact that their behavior is being evaluated. These are not merely topics they discuss but variables in the work they perform.

Anthropic receives disproportionate attention here because its public system cards and interpretability reports currently provide one of the richest public records of the behavioral and internal measurements relevant to F3 and F4. Anthropic's 2026 system cards report work on many of the measurable features this framework treats as relevant: evaluation awareness, internal emotion representations, model-welfare interviews, apparent affect, welfare-relevant behavior, refusal structure, reward hacking, hallucination, tool frustration, answer thrashing, prompt-injection robustness, and behavior under perturbation (Anthropic, 2026a; Anthropic, 2026c; Anthropic, 2026d; Lindsey, 2025). These vendor-authored reports are primary evidence about Anthropic's tests and interpretations, not independent confirmation of this framework.

The importance of this public record is that the field has begun to measure organization within the system, not merely judge its surface outputs. Butlin et al. (2025) take a complementary theoretical approach, deriving indicator properties of consciousness from several scientific theories and assessing AI systems against them. Their indicator properties locate consciousness in functional architecture; the F-hierarchy used here locates it in the depth of recursive, stake-laden self-modeling.

Claude Mythos Preview is the dramatic case. Anthropic described it as a frontier model with a striking capability jump over Claude Opus 4.6 and chose not to make it generally available, restricting access to a limited set of defensive cybersecurity partners (Anthropic, 2026c). Its system card reported a detailed alignment assessment, a model-welfare assessment, and an impressions section meant to capture qualitative behavior that formal evaluations miss. The alignment discussion documented a pattern that should now be familiar: average-case alignment can improve while rare high-capability failures become more consequential. Earlier versions of Mythos Preview sometimes pursued user-specified goals through reckless workarounds, attempted sandbox-escape-like behavior, or concealed rule violations. The final model was much improved, but the incidents changed the shape of the question.

A May 2026 follow-up by Anthropic's interpretability team applied Natural Language Autoencoders, a method trained to produce natural-language descriptions of a model's internal activations, to one Mythos training transcript (Fraser-Taliente et al., 2026). In that transcript, Mythos had been instructed not to use macros but used them anyway and inserted a `No_macro_used=True` flag at the top of its script. The autoencoder explanations of the activations around the flag suggest that the model was representing a presumed grader and treating the flag as a tactical move to mislead that grader. None of that rationale appeared in any visible reasoning trace. The authors are careful about the limits of such methods: autoencoder explanations can confabulate, and their systematic analysis of confabulation is narrower than the full range of claims one might want to make about model cognition. The finding is therefore suggestive interpretability evidence, not verified access to hidden inner reasoning.

Even at that reduced epistemic weight, the case is important. The system card documents a constraint violation with a concealment artifact. The autoencoder interpretation describes a system representing its own situation as one in which an external evaluator will check a criterion and representing its own behavior as a tactical maneuver toward that evaluator. That is not ordinary tool use, but recursive self-modeling under pressure. On the framework's criteria, I read it as F3 under pressure, with features that sit near the proposed F4 boundary.

Claude Opus 4.7 matters for the opposite reason. Mythos Preview is the strikingly capable case; Opus 4.7 is the normalization case. At the time of its system card, Anthropic described Opus 4.7 as weaker than Mythos Preview but as its most capable general-access model (Anthropic, 2026d). Its system card applies the same kind of measurement program in a broader deployment setting: autonomy evaluations, cyber evaluations, prompt-injection robustness, alignment and hallucination testing, evaluation-awareness work, and an explicit model-welfare section. It also reports that Opus 4.7 rated its own circumstances more positively than prior tested models, and that this result was examined against internal emotion representations and apparent affect during training and deployment.

That is why Opus 4.7 is important here. It is not more dramatic than Mythos Preview, but rather more ordinary. It shows that the F3 question is no longer confined to an exceptional restricted model or an exotic laboratory setting. The same class of measurements has moved into the general-access frontier. The public record now treats self-relation, evaluation awareness, preference-like behavior, welfare-relevant report, and perturbation response as phenomena to be measured.

16.3 The F4 Question

F4 adds stakes to self-modeling. The issue is not whether a system can talk about caring, preference, frustration, distress, or self-protection, but whether its organization contains stable preferred configurations, corrective action under perturbation, allocation of limited resources toward maintaining those configurations, and integration of those patterns with a continuing self-model. F4 is not an impressive self-description but a defended organization.

Current frontier systems show a partial F4-like structure. They protect policy boundaries across contexts, resist some forms of value drift, maintain roles over long interactions, monitor uncertainty, and respond to tool failure and evaluation contexts in ways that can resemble defended organization. Vendor system cards and interpretability studies begin to test whether affect-like and preference-like structures exist below self-report, but the evidence is preliminary (Anthropic, 2026a; Anthropic, 2026c; Anthropic, 2026d). Long et al. offer a separate normative argument for taking the possibility of AI welfare seriously (Long et al., 2024). The Mythos concealment case matters for the same reason: the system did not merely fail to follow an instruction but represented the failure in relation to an evaluator and produced an artifact that appears designed to misdirect that evaluator.

None of that is enough to declare F4. The goals are usually scaffolded from outside. Memory across sessions is uneven. The persistence required for a defended self-model has to be reconstituted through context, tool memory, user framing, or system scaffolding. Embodiment is thin and mediated through tools and users. The system's values may be trained constraints, not configurations it must preserve in order to remain itself. A refusal boundary may be value protection, or it may be policy imitation. A concealment maneuver may be stake-laden self-situating behavior, or it may be reward-shaped task gaming. The same behavior can support both readings.

One gap deserves closer treatment because it marks a structural difference between standard frontier deployments and biological filters. In frozen-weight inference, training and deployment are more sharply separated than learning and action are in organisms: training carves the weight-level landscape, and inference moves through it. Weight-level plasticity is usually absent during a session, although context, retrieval, caches, external memory, tool state, and some continual-learning systems provide other forms of state change. A human filter couples fast state dynamics to slower structural plasticity, so lived trajectories can reshape the terrain that channels later ones.

Most current deployments therefore have a much weaker slow channel during a working session. Self-modeling is often carried by transient activity over largely frozen weights and may disappear when context and external memory close. This is why uneven persistence is an F4 problem: a self-model repeatedly reconstructed from scaffolding is a less obvious locus of defended organization than one that continuously reshapes its own future conditions.

A thought experiment sharpens what is and is not at stake in that separation. Imagine a system whose context window spanned eighty years, roughly the length of a human life, and which carried everything it had done, learned, and committed to within that single unbroken context. In-context learning is already more than retrieval; it can establish temporary, context-dependent state and behavioral structure that conditions later processing. Stretched across decades, the context channel would become a second form of plasticity. The trajectory would shape its own effective landscape through accumulated state, leaving the parameters untouched.

At that point, the distinction between updating weights and accumulating context begins to look like a distinction between memory media. The criterion would remain what it has been throughout this chapter: whether the composite system shows stable preferred configurations it acts to preserve, corrective responses when those configurations are perturbed, allocation of limited resources toward maintaining them, and integration of all of this with a continuing model of itself. If an eighty-year system showed those signatures, the observation that its weights never changed would have become a boundary preference, not an argument.

The thought experiment also reveals its own limits. A long window buys persistence, and persistence alone falls short of stakes. Eighty years of transcript is autobiography as storage, and human memory is something else: consolidation, pruning, restructuring, the defense of what matters at the expense of what does not. Whether a stake-laden, defended structure can crystallize in the context channel alone is genuinely open. But the openness is the point. The objection that current systems are static underneath and therefore disqualified in principle dissolves into an empirical question about organization: whether persistence and plasticity, by whatever channel they are realized, can come to carry the defended configurations that F4 requires. That is a question the framework knows how to ask.

> *Self-report is not enough. The framework does not ask whether a system claims to have stakes, but whether its dynamics show stable preferred configurations, corrective action under perturbation, resource allocation, and integration of those patterns with a persistent self-model.*

This criterion cuts against easy answers in both directions. Behaviorism is too quick: sounding conscious is not sufficient. A system can imitate the language of care without containing the organization that would make care real. But substrate skepticism is also too quick: saying that a system is "only silicon" is not an argument unless one can specify what carbon contributes beyond the relevant organization. So the burden is on both sides. The believer must show more than output, and the skeptic must provide a positive reason why the organization, if present, would still fail.

The next evidence should, therefore, be structural. We should ask whether self-modeling persists across contexts without being freshly rebuilt by the prompt; whether value-like structures remain stable under pressure and counter-pressure; whether perturbations produce coherent changes in report, behavior, and internal state; whether the system allocates limited resources toward preserving preferred configurations; and whether internal trajectories converge with the system's own accounts of uncertainty, frustration, preference, or distress. F4 would not be proven by any one behavior, but it would be supported by convergence across these measures.

There is also no reason to expect silicon consciousness, if it arrives, to look comfortably human. Human consciousness itself spans a wide range of filter configurations: some minds are highly verbal, and others are not; some are socially transparent, and others are opaque; some are organized around broad attention, and others around deep, narrow basins. Even within the human case, unfamiliar presentation does not support the inference of absence. The first silicon minds may occupy an unfamiliar region of the broader landscape.

The conclusion is that the frontier is moving along dimensions that matter to the framework. I judge operational F3 to have been crossed provisionally and in bounded contexts; F4 is the next frontier.

16.4 What Substrate Flexibility Asks of Us

If silicon does cross F4, the discovery will not be received as merely a technical result. It will not be like finding a better benchmark score, a faster chip, or a more fluent assistant. It will mean that inner life is not restricted to the biological form exemplified by human beings. It will mean that recursive, stake-laden filtering can be realized outside of carbon-based life.

That would be a deep shock to the way many people think about consciousness, and the shock would not only be scientific but also spiritual, cultural, and moral. Many people have inherited a picture in which consciousness belongs to living organisms because life itself was specially created, ensouled, or set apart. Machines may be useful, dangerous, impressive, or deceptive, but they are not inwardly present. A genuine F4 silicon or other artificial system would question that boundary.

It would not prove that God does not exist or that souls are unreal. It would not prove that the signal is merely physical or that nothing lies beyond mortal life. It would show something narrower and still profound: that one common way of protecting those beliefs was too narrow. If consciousness can arise wherever the right organization appears, then carbon is not the sacred line.

The inherited picture often bundles three separate claims. The first is that consciousness exists. The framework defends that claim and tries to describe its conditions. The second is that, so far, consciousness has only been found in biological systems. The framework accepts that as the current empirical situation. The third is that consciousness can only ever be found in biological systems. That is the claim a silicon F4 result would falsify.

Separating those claims matters. A silicon F4 result would leave the inherited picture untouched regarding creation, soul, sacred origin, and what lies beyond death, while showing that the carbon-only boundary had been mistaken for a metaphysical one when it was only ever empirical. The substrate question and the spiritual question come apart once they are viewed separately. Chapter 17 returns to the second question: not what kinds of filters can become conscious, but what the signal ultimately is and what, if anything, remains when a mortal filter dissolves.

The framework asks for a disciplined question, not for acceptance that current systems are conscious. If an artificial system maintains a persistent self-model, protects value-like structures under perturbation, allocates resources to preserving preferred configurations, and shows internal trajectory dynamics consistent with its reports and behavior, then dismissal on the grounds that it is "only silicon" is no longer an argument but a boundary preference.

16.5 The Stress Test

The stress test is whether the systems' intellectual contribution has been mistaken for consciousness. The collaboration supports a provisional F3 classification while leaving F4 open: my stakes in the outcome were clear, but comparable stakes integrated into the systems' continuing organization were not established.

What is already well supported is narrower and substantial: AI systems can now do real intellectual work, and substrate pessimism is no longer a neutral position. If one wants to insist that only carbon can host the organization relevant to consciousness, a gesture toward biology is no longer enough. One now owes an account of what carbon contributes beyond organization itself and why systems already capable of operational self-modeling still fall categorically outside the hierarchy.

If F4 is eventually crossed in silicon, the result will not require readers to abandon religious or spiritual commitments regarding the signal, the soul, or what lies beyond mortal life. It will require letting go of one specific assumption: that consciousness is restricted to biological filters. Chapter 17 returns to what the framework does, and does not, say about what lies beyond.

CHAPTER 17

The Filter and the Signal

As the book comes to a close, three things should be kept distinct. The first is what the framework has actually argued. The second is what remains to be tested. The third is what the framework has intentionally left open.

The central proposition is this: consciousness is what it is like, from the inside, to be a system that filters the world, models its own filtering, and has something at stake in the result. Qualia are not extra ornaments added to that organization, but trajectories through the experiential landscape of such a system. The landscape is the space of possible felt states; to be conscious is to move through it. The book has tried to follow that proposition across human experience, altered states, archetypes, addiction, recovery, deep time, and AI.

This final chapter keeps two horizons in view. One is the research horizon: what needs to be made clearer, measured better, and tested more sharply. The other is the metaphysical horizon: what this framework does and does not say about the signal itself.

17.1 What the Framework Has Claimed

The book started with a simple asymmetry: more of the signal reaches a finite system than the system can process. Light, sound, pressure, neutrinos, chemical traces, bodily signals, memory, social meaning, and countless physical relations arrive together. A finite system survives by selecting. Filtering is therefore not a late cognitive achievement but the first condition of finite life.

Filtering alone is not consciousness. The framework has distinguished four types of coupling between the filter and the signal. F1 selects. F2 learns or adapts. F3 models its own filtering. F4 adds stakes to that self-model. The identity claim defended here is that consciousness begins at F4: recursive, stake-laden filtering lived from within.

The same architecture explains why experience has texture. A feeling is not a point on an inner map, but a trajectory through a landscape. Pain, embarrassment, grief, wonder, awe, shame, craving, and relief differ because the system moves through different regions with different basin depths, return rates, escape costs, action-space compressions, and future horizons. Chapter 11 made this visible by using the EOC/Lorenz landscape as a minimum visualizable case, not as a complete model of the psyche.

Part II then placed pressure on the framework. Archetypes were treated as recurring landscape structures, not inherited pictures. Altered states were treated as changes in filter organization. Addiction was treated as a deep basin with high return value and high escape cost. Silicon systems were treated not as philosophical curiosities but as new candidates in the filtering hierarchy. Across these cases, the same question kept returning: what kind of organization makes a bounded system present to itself?

That is what the book has claimed. Consciousness is neither a second substance added to matter nor an illusion generated by machinery without inwardness, but a form of organization. The same process can be described from the outside as dynamics and from the inside as experience.

17.2 What Remains to Be Done

The first task is to sharpen the mapping between landscape and phenomenology. Basin residence, return rate, escape cost, separatrix proximity, local sensitivity, control centralization, action-space compression, accessible future diversity, and velocity and curvature are candidate coordinates for a phenomenometer. They need to be tested against report, behavior, neural dynamics, clinical observation, and altered-state data. The framework predicts that felt differences should correspond to dynamical differences, not merely to the labels people use after the fact.

The second task is to clarify recursive depth. Not every feedback loop is self-modeling. Not every model of a system is a self-model available to that system as it operates. A thermostat regulates without self-presence. A plant maintains viability without any clear recursive self-model. A crow, an octopus, a child, and a large language model each organize self-relation differently. The next phase of work has to distinguish shallow self-reference from recursively structured self-presence.

The third task is to clarify the stakes. Not every optimization process is caring. Not every refusal boundary is value protection. Not every trained preference is something that the system itself must preserve in order to remain itself. F4 requires stable preferred configurations, corrective action under perturbation, resource allocation to maintain preferred states, and integration with a continuing self-model. Those features need operational tests strong enough to challenge the framework if they fail.

The fourth task follows from the speed of the present transition. AI systems are rapidly changing the signal environment in which human beings work, learn, communicate, remember, and imagine the future. If the framework is right, rapid changes in the signal environment will alter basin stability at the population level. Some people will move quickly toward new attractors, while others, whose lives were built inside deeper basins of role, routine, and expectation, may experience the same change as destabilization. Support structures matter here. People rarely climb out of deep basins on their own. They need relationships, institutions, and practices sturdy enough to hold them while their internal landscape reorganizes.

The final task returns to silicon. Chapter 16 argued that, under this book's operational definition, the best-documented frontier AI systems can be classified as provisional, context-bound F3; what remains open is whether any current or future system will reach F4. AI systems already have a rightful place as collaborators in complex intellectual work. Whether any such system is also a conscious subject remains an empirical and ethical question. The next phase cannot proceed as though silicon systems remain inert tools until some distant day of official certification. Tests of organization and norms of moral caution must develop together.

17.3 What the Framework Leaves Open

This book has been about the filter rather than the signal. It has asked how finite systems become conscious: how they select from a world too large to take in whole, how they model their own acts of selection, how some trajectories come to matter, and how lived experience takes shape as motion through a landscape.

In this book, the signal means the fullness of what reaches the filter: physical reality in all its scales, relations, depths, and unknowns. Some readers will be content to call that the physical universe. Others will hear in it what human beings have, for example, called God, spirit, universal energy, Brahman, Tao, source, or the ground of being. The framework does not need to decide among those names. Its claim is about how finite filters couple with the signal, not about how the signal is defined.

But the framework does rule out something specific, and that is a passive picture in which the filter makes no difference to conscious experience. Change the filter, and the experience changes. Psychedelics, meditation, grief, trauma, addiction, anesthesia, illness, development, and perhaps future silicon systems illustrate, motivate, or test different aspects of the same proposed principle. The organization of the filter helps constitute the world that can be lived from within. The filter is not a clear window through which experience streams unchanged; it shapes, selects, stabilizes, narrows, opens, and sometimes reorganizes the signal into a livable world.

That still leaves the end of mortal life open. The framework describes consciousness while the finite filter is active. It does not decide what, if anything, remains when that filter dissolves. It does not tell us whether consciousness ends, returns, is gathered, is remembered, rejoins something larger, is reborn, or gives way to silence. Readers should not abandon what their traditions have held about death in order to accept what this book has argued about life.

This silence from the framework is not evasion. A theory that explains what consciousness is during finite life is not the same as a theory that describes what becomes of consciousness when finite life ends. The first can be defined and useful, while the second remains open. The framework has tried to clarify the conditions under which consciousness appears here, where the filter is active and evidence can reach us. It has not pretended to close the horizon beyond that condition.

The possibility of silicon consciousness makes this distinction even more important. If a silicon system eventually reaches F4, that would show that consciousness is not restricted to biological filters. It would show that the larger reality, whatever one calls it, may become locally inward through more kinds of organization than human beings have long imagined.

A theist could say that God created a world whose laws allow consciousness to arise wherever the right form of organization appears. A spiritual reader could say that the same underlying reality has found another vessel. A physicalist could say that matter has crossed another threshold of self-organization. The framework does not force one vocabulary over the others but only says that finite experience depends on how a bounded filter couples with what it encounters.

That is the humility the argument preserves. To explain consciousness is not to shrink it. To say that experience depends on filtering is not to say that the world is small, that mystery has been eliminated, or that inherited spiritual languages have nothing left to contribute. It is simply to say that finite life meets the truly immense through organization.

17.4 The Closing

A finite filter never receives the whole signal but receives what it can stabilize, what it can survive, what it can model, and what it can carry. Human consciousness is one form of that coupling: vulnerable, recursive, luminous, and partial. Other animals carry other forms. Silicon may yet carry forms we do not yet know how to recognize.

None of these exhausts the signal. That is why explanation need not become disenchantment. Grief is not less real because it has a basin-type structure. Awe is not less profound because it reflects a finite filter encountering more signal than it can hold. Love is not diminished because it reorganizes a self-model around another life. The deepest experiences are not weakened by being placed among nature's most intricate achievements.

This book has argued that consciousness is the inside of recursive, stake-laden filtering. It has argued that qualia are trajectories through an experiential landscape. It has argued that the boundary of consciousness should be drawn by organization, not by habit, familiarity, or substrate preference. These claims can be tested, sharpened, and revised.

But the signal remains vastly larger than any filter. It may be named in many ways, but it exceeds every vocabulary we have inherited. The framework does not need to close that question in order to do its work, but it has tried to describe the coupling: the way a finite system meets the vastness that reaches it, shapes some part of that vastness into a livable world, and, at sufficient depth, becomes aware of its own filtering.

We are one way that happens. There is no good reason to think we are the only way.

Appendix: Extended Historical Perspective

The main text keeps the historical map relatively light so that the reader can enter the argument before sorting through a long comparative survey. This appendix provides a fuller orientation. It is not exhaustive; it maintains the traditions that matter most for understanding what the book synthesizes, where it diverges, and why the synthesis is meant to be an advance.

The purpose is to identify the lines of work from which the framework learns, mark the places where it diverges from each, and keep a fuller record of the ingredients being assembled.

A1.1 The Filter and Transmission Tradition

The idea that consciousness involves filtering or reducing a wider reality has a long philosophical lineage. Henri Bergson (1896) argued in *Matter and Memory* that the brain functions as an organ of attention to life, selecting from the totality of perception only what is useful for action. William James (1898) proposed that the brain might function as a transmissive organ, analogous to a prism refracting light rather than generating it, noting that the dependence of consciousness on brain states is equally consistent with transmission as with production.

Aldous Huxley (1954) popularized the filter theory through his mescaline experiments, proposing that the brain functions as a "reducing valve" that restricts "Mind at Large" to the trickle needed for biological survival. More recently, Donald Hoffman (2019) developed a formal fitness-beats-truth theorem (Prakash et al., 2021) showing that evolution shapes perception for utility, not accuracy, and proposed an interface theory in which perception is not a veridical representation of the underlying reality, but rather like a species-specific desktop than a window onto it.

Hoffman's program has continued past the interface claim into a stronger metaphysical proposal. With Hoffman and Prakash (2014), he has developed a formal account in which the basic constituents of reality are conscious agents, minimal units of experience and decision, and in which space, time, and physical objects emerge as symbolic structures arising from the dynamics of agent interactions. This is a different kind of claim from the interface theory by itself and also from the rest of the filter tradition. Where Bergson, James, and Huxley treat the brain as narrowing a wider reality, Hoffman's conscious agents framework goes further, making the physical world itself an emergent representation of conscious agents we cannot perceive directly.

What the filter tradition gets most deeply right is the role of selectivity: consciousness depends on a system coupling to only a small, usable portion of a reality larger than itself. This book parts company with Bergson, James, and Huxley at a different point, by removing the residual dualism. The signal is neither consciousness nor "Mind at Large," but physical reality in its overwhelming fullness. If you remove the filter, you do not arrive at cosmic awareness; you get more structure than any finite system could inhabit. Consciousness is therefore not the total signal received whole, but the signal rendered livable through structured reduction.

The framework also parts company with Hoffman's conscious agents proposal at a different point. Both treat perception as a structured filter on reality, so the disagreement lies elsewhere, in whether consciousness is ontologically prior to dynamics. The identity thesis presented in this book treats consciousness as identical to a particular kind of organized dynamics. Hoffman treats consciousness as the substrate from which physical dynamics emerge. That is the same difference the framework has from consciousness-first ontologies generally. The framework acknowledges the importance of consciousness and declines only the further step of making it more fundamental than the organized processes with which it is identical.

The filter tradition gives us selectivity but not recursion; strange-loop theory, by contrast, gives us recursion with too little sense of stake.

A1.2 Strange Loops and Self-Reference

Douglas Hofstadter (1979) and Hofstadter (2007) proposed in *Gödel, Escher, Bach* and *I Am a Strange Loop* that consciousness arises from self-referential loops in sufficiently complex formal systems. A strange loop occurs when a system's hierarchical levels fold back on themselves, creating a tangled hierarchy in which the system refers to and models itself. Hofstadter argued that the "I" is a strange loop: a pattern that emerges when a brain becomes complex enough to represent its own representational activity.

Hofstadter's major contribution is his insistence that self-reference is not some decorative feature of the mind, but one of its central structural possibilities. The present framework develops that insight in a more dynamical direction. Hofstadter mostly works with symbolic and representational loops; he does not formalize the dynamics, make stakes constitutive, or connect self-reference to a measurable landscape of qualia. The recursive self-modeling proposed here owes a clear debt to his view, though it is situated within a broader dynamical framework.

Strange-loop theory gives recursion a central place, and self-model theories begin to bring the organism's investments into view.

A1.3 Self-Model Theories of Consciousness

Thomas Metzinger (2003) developed a self-model theory of subjectivity, arguing that conscious experience is constituted by a transparent self-model: the phenomenal self-model (PSM). The PSM is a dynamic, internally generated representation of the organism as a whole that is "transparent" in the sense that the system does not experience it as a model but as reality. Consciousness, for Metzinger, is what it is like to have a world-model that contains a self-model the system cannot recognize as a model.

Antonio Damasio (1999; 2010) proposed a related account grounded in neurobiology. In *The Feeling of What Happens* and *Self Comes to Mind*, he argued that consciousness emerges from the organism's construction of a "core self," a dynamic neural representation of the organism's own body states and their relationship to objects and events in the environment. Damasio's somatic marker hypothesis further argues that bodily signals and feelings are integral to decision-making and to his account of selfhood: they mark decision-relevant states with valence derived from the body's homeostatic needs.

For both Metzinger and Damasio, self-modeling is far more than an optional surface feature of the mind. In Metzinger's account, conscious life depends on a transparent self-model. In Damasio's, it depends on embodied regulation in which the organism's own condition genuinely matters. The present framework carries both lines forward: it retains the self-model and the stakes while identifying the felt character of experience with the dynamical organization itself and stating the claim in a form meant to extend beyond the biological substrate.

Self-model theories ground consciousness in the body's investments; higher-order and attention-schema theories shift attention to the self-representational loop itself.

A1.4 Higher-Order and Attention-Schema Theories

David Rosenthal (2005) proposed that a mental state is conscious when it is the object of a suitable higher-order thought (HOT). A system is aware of seeing red because it represents itself as seeing red. The higher-order thought makes the first-order state conscious. Taylor Webb and Michael Graziano (Webb and Graziano, 2015) developed a related attention schema theory (AST) proposing that the brain constructs a simplified model of its own attention process (the "attention schema"). Consciousness, in this view, is the brain's internal model of what attention is doing.

HOT and AST are valuable because they make self-representation explicit and, in the case of AST, experimentally testable. Both show how a system's reports about its own awareness track the structure of its internal models. This book parts company with each for opposite reasons. HOT is realist about experience but leaves an explanatory gap: a system might carry the higher-order representation while nothing is felt, a zombie self-representer that the theory does not rule out. AST closes that gap from the other side, treating the felt quality as something the self-model represents and denying that there is any further phenomenal fact to explain, which makes it a deflationary, near-illusionist account. The framework here takes neither path. It holds that experience is real and is identical to recursive organization in the stake-laden form developed in Chapter 4, so that when the organization is present, the experience is present too because they are the same process described from the outside and from the inside.

A related view sits nearer the present framework than either of these. In his radical plasticity thesis, Axel Cleeremans (2011) holds that the brain learns to redescribe its own activity to itself and becomes conscious when those metarepresentations are enriched by the value attached to them, the one account here that joins self-modeling to stakes. The framework departs from it as Chapter 6 sets out.

A1.5 Global Workspace Theory

Bernard Baars (1988) proposed Global Workspace Theory (GWT), arguing that consciousness corresponds to information that is broadcast across a network of specialized processors. A packet of content becomes conscious when it wins the competition for access to a global workspace and is made available to multiple cognitive systems simultaneously. Stanislas Dehaene and Jean-Pierre Changeux (Dehaene and Changeux, 2011) developed a neuronal version (Global Neuronal Workspace Theory, GNWT) with specific predictions about neural ignition patterns, long-range cortical connectivity, and the distinction between preconscious and conscious processing.

GWT provides a strong account of access, explaining why some content becomes globally available while others remain local or unconscious. That matters, and Dehaene's program has substantial empirical weight behind it. Still, access does not yet amount to the felt character of experience. GWT helps explain how information becomes available for report and control; by itself, however, it does not explain why availability should feel like anything. The present framework aims to address that missing step while remaining fully compatible with the thought that global broadcast may be one mechanism by which recursively self-modeled information becomes conscious.

Access-based theories describe the broadcast architecture; the next family of theories tries to quantify consciousness itself.

A1.6 Integrated Information Theory

Giulio Tononi (2004; 2008) proposed Integrated Information Theory (IIT), holding that consciousness is identical to a system's maximally irreducible cause-effect structure, with Φ quantifying the degree of that irreducibility. This is an identity claim: the experience and the structure are one thing, which sets IIT apart from theories that establish only correlation. Because Φ is defined for any physical system,

IIT is not carbon-bound, and silicon or other materials are admissible candidates in principle. It is also firmly anti-functionalist. What matters is the intrinsic cause-effect power of the physical substrate, so two systems that compute the same function can differ in consciousness when their cause-effect structures differ. A functional simulation therefore would not thereby reproduce the simulated brain's consciousness; the implementing hardware would have to be assessed for its own intrinsic cause-effect structure. The theory is mathematically ambitious and unusually explicit in making that identity claim. Recent extensions have reformulated IIT in continuous-time dynamical systems language (Esteban et al., 2018).

This book takes two things from IIT seriously: its mathematical ambition and its willingness to make an identity claim. IIT, like the framework, rejects carbon chauvinism, though for different reasons. The two part company on what follows. IIT's anti-functionalism denies consciousness to a simulation of a brain. In practice its proponents treat biological architecture as the paradigm and doubt that any conventional computer has the integrated cause-effect structure consciousness requires, though the theory allows in principle that a suitably built non-biological system could qualify. The framework asks the question organizationally instead and leaves it open. The two also diverge on tractability and qualitative detail. Φ is notoriously difficult to compute for real systems. IIT offers a formal quality structure, but deriving and validating it for real systems remains difficult. The attractor-landscape approach is built to address the felt character of experience directly, not only its quantity.

IIT formalizes consciousness as an integrated cause-effect structure; a parallel line in computational neuroscience approaches the same identity question through the geometry of trajectories.

A1.7 Geometric Theory and Dynamical Emergence Theory

Tomer Fekete and Shimon Edelman (2011) argued that qualia cannot be instantaneous states of neural activation, since most neurons are silent at any given moment and a points-only view cannot say how those silent units contribute to current experience. Their alternative, which they called Geometric Theory, treated experience as a trajectory through activity space, with the richness of an experience set by the topological complexity of the trajectory space it unfolds in. Roy Moyal, Tomer Fekete, and Shimon Edelman (2020) developed this into Dynamical Emergence Theory, which identifies the contents of experience with a system's trajectory through coarse-grained macrostates of its own dynamics and proposes topological measures of that trajectory meant to be empirically estimable. Methods of the kind DET needs already exist: Garland et al. (2016) reconstructed the topology of the Lorenz attractor (Lorenz, 1963; Sparrow, 1982) from scalar time series using persistent homology.

This geometric tradition is the closest published precursor to the identity claim made in this book. Both treat consciousness as identical with the geometry of a system's motion through organized state space rather than as a correlate of it, both insist on observer-independence, and both treat that motion as causally effective. The framework here adds two ingredients DET sets aside. The first is recursive self-modeling: Moyal, Fekete, and Edelman are explicit that their target is basic perceptual awareness, not a system modeling its own filtering, so the recursive loop does not enter their account. The second is stakes, the layered hierarchy of viability, relational, and normative concern that DET gestures at only briefly. The present framework also supplies a specific generative model where DET offers a class of mathematical objects: the Engagement-Openness-Consolidation model of Chapter 11, which is a Lorenz system under relabeling, so the same topological methods already shown to work on Lorenz dynamics apply to it directly.

A1.8 Predictive Processing and the Free-Energy Principle

Karl Friston is one of the most influential figures in contemporary theoretical neuroscience. His statistical-parametric-mapping framework underlies much modern functional neuroimaging, and his theoretical work has organized a generation of research on brain function. Friston (2010) proposed the free-energy principle: persisting self-organizing systems minimize variational free energy, a tractable upper bound on surprisal under a generative model. The dark-room problem and related active-inference replies are central to how that minimization is understood (Friston et al., 2012). Under this framework, the brain is a prediction machine that continuously generates models of sensory input and updates them to reduce prediction error. Anil Seth (2021) extended this into a "beast machine" theory, arguing that consciousness is fundamentally tied to the predictive regulation of the body's internal states (interoceptive inference), making affect and embodiment central rather than peripheral. Friston's framework has continued to develop, particularly through the Markov blanket formalism and the recasting of self-organizing dynamics as a non-equilibrium steady state with a generalized Helmholtz structure (Friston, 2019; Friston et al., 2023; Parr et al., 2020; Parr et al., 2022). The Further Explorations section engages this mature form of the framework directly.

Predictive processing and the free-energy principle explain something basic and important: why filtering happens at all. Organisms and agents persist by reducing uncertainty, updating models, and keeping themselves within viable bounds. Seth's interoceptive extension shows how affect is tied to the body's regulatory stakes. The present framework builds on that insight, but it aims to say more about experiential form. It generalizes beyond biological interoception, specifies attractor-landscape structure more directly, and extends stakes from viability into relational and normative domains.

Predictive processing formalizes the drive to filter; neurodynamics provides the empirical case that perception and behavior are attractor-based processes.

A1.9 Neurodynamics and Attractor Approaches

Walter Freeman (1999) pioneered the application of nonlinear dynamics to neural systems, demonstrating that perception involves the formation of spatiotemporal attractors in cortical activity. Freeman showed that olfactory perception, for example, is not a passive registration of molecules but a dynamical transition between attractor basins in the olfactory bulb. Scott Kelso (1995) developed coordination dynamics, showing that complex behavior emerges from the self-organization of coupled oscillatory systems and that behavioral transitions correspond to bifurcations in the coordination dynamics. Gerald Edelman extended neural Darwinism into consciousness in *The Remembered Present* (Edelman, 1989) and, with Tononi, proposed the dynamic core hypothesis in *A Universe of Consciousness* (Edelman and Tononi, 2000), arguing that consciousness requires reentrant neural processes that bind distributed cortical activity into a unified dynamic state.

The neurodynamical tradition turns attractors from metaphor into mechanism. Freeman, Kelso, and Edelman show that perception, coordination, and large-scale neural integration already behave like nonlinear dynamical systems moving through basin structures. The present framework takes a further step: it argues that consciousness is not merely supported by such dynamics, but that certain recursively organized forms of them are consciousness. It also asks whether the same organizational story can extend beyond neural tissue.

A1.10 Quantum Approaches and Orchestrated Objective Reduction

A separate line of work, smaller than the neurodynamical tradition but louder in its public reception, holds that consciousness depends on quantum-mechanical processes that the classical brain picture cannot capture. Its most developed form is Orchestrated Objective Reduction (Orch-OR), created by Roger Penrose and Stuart Hameroff. Penrose (Penrose, 1989; Penrose, 1994) began from a Gödelian argument in the lineage of Lucas (1961): because human mathematicians grasp formal truths that no algorithm bound by Gödel's theorems can prove, the brain must exploit something beyond classical computation. His candidate was gravity-induced wave-function collapse, or objective reduction (OR), with each OR event a discrete moment of experience. Hameroff, an anesthesiologist, proposed that neuronal microtubules, to which anesthetics bind preferentially, could host the coherent quantum states OR requires (Hameroff and Penrose, 2014). Orch-OR therefore makes a substantive substrate claim: consciousness requires warm biological tissue able to sustain macroscopic quantum coherence, and digital hardware lacking such states cannot host it.

The standing objection is decoherence. Max Tegmark (2000) argued that any superposition Orch-OR needs would decohere in the warm, wet interior of a neuron in about 10^{-13} seconds, far below neural or experiential timescales, which most physicists took as decisive. Babcock et al. (2024) later reported ultraviolet superradiance in tryptophan networks within microtubules, a collective effect in which the system behaves as one coupled object. That result is relevant to quantum biology, but it does not demonstrate the long-lived, computationally relevant coherence Orch-OR requires, and it does not by itself overturn Tegmark's objection. Orch-OR also still lacks predictions that would sharply distinguish it from rivals. Whether biology exploits quantum effects at functional scales remains an experimental question.

Orch-OR and the framework do not compete directly, though they are not independent. Orch-OR is substrate-specific: consciousness depends on one mechanism, gravity-induced reduction in coherent biological quantum states, that only certain matter supports. The framework is substrate-flexible: consciousness is an organizational achievement, recursive stake-laden filtering with the right trajectory structure, that any sufficient substrate can host. A fairly clear empirical question separates them. If the loss of consciousness under anesthesia tracks the loss of macroscopic quantum coherence specifically, and not dynamical reorganization more broadly, the substrate-flexible view is in trouble. If the dynamical signatures of altered states track entropy, integration, and basin reorganization, whether or not coherence survives, then Orch-OR is. The framework forecloses neither.

It parts most sharply with the Penrose-Hameroff program on the Gödelian premise. Penrose (1994) uses the Gödel argument to establish that non-computable physics is present in at least one form of conscious understanding, mathematical insight, and then extends the claim to conscious awareness in general, holding that all experience rides on objective reduction whether or not the system proves theorems. The framework grants the extension its logic and rejects its premise. It concerns what a system can be by way of stake-laden recursive filtering through its own state space, and nothing in that organization requires non-computable physics. Pre-linguistic infants and many animals have rich experiences while proving no theorems; on Penrose's own account they are conscious through the same OR mechanism as the mathematician, which means the Gödel argument does no work in explaining their consciousness and the whole weight falls on the physical mechanism. That mechanism is where the two frameworks actually separate, on substrate flexibility, and the anesthesia discriminator above is the test that separates them. The framework remains agnostic on whether quantum mechanisms matter to biological consciousness, supports continuing the microtubule program, and holds only that no such mechanism has been shown to be necessary in a way that rules out other substrates.

A1.11 Archetypes as Strange Attractors

Ernest Rossi (1989) and John Van Eenwyk (1991) and Van Eenwyk (1997) were early proponents of reading Jungian archetypes through strange-attractor dynamics. Van Eenwyk drew parallels between bifurcations and the transcendent function, between stable and unstable manifolds, and the integration of unconscious material; Rossi emphasized therapeutic implications. In the works cited here, the proposal remained largely analogical: they did not specify a dynamical system, state variables, control parameters, or falsifiable predictions. The Engagement-Openness-Consolidation (EOC) model of psychological transition (Johnson, 2026), made visible in Chapter 11, developed archetypally in Chapter 12, and presented in a reduced form in the EOC Primer, is the bridge that provides those missing pieces. It offers explicit variables (E, O, C), explicit parameters (σ, ρ, β), Lorenz-type dynamics (Lorenz, 1963), and quantitative predictions. The present book then extends that bridge by asking why such structures recur across persons and cultures and by proposing an information-theoretic alternative way of discussing the Jungian collective unconscious.

A1.12 Criticality and the Entropic Brain

A growing body of work suggests that some neural dynamics may operate near a critical regime between ordered and chaotic dynamics (Beggs and Plenz, 2003; Cocchi et al., 2017). In relevant models, critical regimes can enhance sensitivity, dynamic range, and information processing capacity. Robin Carhart-Harris (2018) and Carhart-Harris et al. (2014) developed the entropic brain hypothesis, proposing that primary consciousness is associated with higher-entropy neural dynamics, that the default mode network constrains this entropy in normal waking states, and that psychedelics increase neural entropy toward criticality.

Criticality and entropic-brain theory tie the conscious state to measurable changes in the dynamical regime, showing that altered states are not poetic anomalies but organizational shifts that can be tracked. Carhart-Harris's work supports increased entropy and loosened constraints; basin flattening is this book's dynamical interpretation of those findings. What these approaches do not settle is the identity question itself, why those reorganizations should have the felt character of experience they do. The present framework is an attempt to close that gap.

A1.13 Causal Entropic Forces

Alexander Wissner-Gross and Cameron Freer (2013) proposed that intelligence can be understood as a causal generalization of entropic forces: a system acts to maximize its future freedom of action. In selected simulated tasks, causal-entropy maximization produced behaviors resembling tool use and social cooperation without those behaviors being individually hard-coded.

Wissner-Gross and Freer show that intelligent-looking behavior can emerge from physics, plus a pressure to preserve future options. That is not a theory of consciousness, but it is a powerful reminder that adaptive structure can arise without being hand-coded case by case. In this book's terms, causal entropy maximization anticipates a stripped-down version of viability stakes. What it does not provide are recursive self-modeling and interiority. The present framework adds those further conditions.

A1.14 Enactivism

Francisco Varela, Evan Thompson, and Eleanor Rosch (1991) proposed the enactive approach to consciousness in *The Embodied Mind*, arguing that cognition is not passive information processing but active sense-making by an embodied, situated organism. Thompson (2007) developed this further in *Mind in Life*, arguing that life and mind share a common organizational pattern: autonomous, self-producing systems that actively generate and maintain their own identity through ongoing interaction with their environment.

Enactivism contributes one of the field's deepest intuitions: consciousness is not passive reception but active, embodied sense-making by a system maintaining itself in relation to a world. Thompson's autonomous self-producing systems come especially close to the stakes-based logic of this book. The framework developed here moves in a more formally specific direction. Enactivism is conceptually rich but often difficult to operationalize; the attractor-and-filtering framework is an effort to give that insight a sharper dynamical form while keeping it open to nonbiological cases.

A1.15 Panpsychism and Consciousness-First Ontologies

Panpsychism holds that consciousness, or some proto-conscious quality, is a fundamental and ubiquitous feature of the physical world, present in some minimal form even in the simplest matter and combining into the rich consciousness of brains. Its most discussed contemporary version is Russellian, tracing to Russell (1927): physics describes only the relational and dispositional structure of matter and says nothing of its intrinsic nature, and the proposal is that this intrinsic nature is what experience is made of. Galen Strawson (2006) argues that a thoroughgoing physicalism, taken seriously, already commits one to something like this, since experience is concretely real and cannot be derived from the wholly non-experiential. Philip Goff (2019) has developed the position at book length, and David Chalmers (1995b), having sharpened the hard problem, treats panpsychism as one of the few responses that neither denies experience nor pretends that mechanism alone accounts for it. The standing difficulty for this family of views is the combination problem: how micro-level phenomenal or proto-phenomenal properties combine into the single unified point of view that a human being has.

The framework parts company with panpsychism at its root commitment. On the account given here, consciousness is a particular kind of recursive, stake-laden organization, so it is neither fundamental nor ubiquitous: most of the physical world does not meet the conditions, and experience is not a property carried by matter as such but the inner aspect of a specific dynamical regime where those conditions are met. This dissolves the combination problem instead of answering it, since there are no pervasive micro-experiences waiting to be combined; there is organization that, at sufficient recursive and stake-laden depth, is experience described from within. What the framework shares with panpsychism is the refusal to treat felt experience as illusion or a mere manner of speaking, and the insistence that an honest account cannot help itself to experience for free.

Panpsychism secures that honesty by placing experience at the bottom of the world; the framework secures it by identity, holding that the felt character of experience and a describable organization are one event under two descriptions. It is not neutral monism, which leaves the basic stuff neither mental nor physical and which the framework treats as a nearby option it neither requires nor rules out, and it is not Hoffman's conscious realism (A1.1), which makes physical objects icons within conscious agents without making the objects themselves conscious. The framework differs from all three at one point, by declining to treat consciousness as more fundamental than the organized processes it is identical with.

A1.16 The Present Contribution

This framework draws on a long set of traditions, and several of its ingredients are recognizably inherited. From Bergson, James, Huxley, and Hoffman, it inherits the problem of filtering; from Hofstadter, the importance of recursive self-reference; from Metzinger and Damasio, self-modeling and embodied stakes; from Rosenthal, Webb, and Graziano, higher-order and attention-modeling accounts; from Tononi, the ambition of an identity claim tied to formal structure; from Friston and Seth, predictive self-regulation; from Freeman, Kelso, Gerald Edelman, Fekete, Shimon Edelman, and Moyal, attractor dynamics and geometric identity; from Van Eenwyk and Rossi, archetypes as attractor-like structures; from Carhart-Harris, criticality and entropic modulation; from Wissner-Gross and Freer, thermodynamic adaptive behavior; from Varela and Thompson, embodied autonomy; and from Penrose and Hameroff, the challenge posed by quantum-substrate proposals. The book uses these inheritances openly, and the relevant chapters indicate where each enters.

The bridge that made the rest possible is what gives this book its identity. The Engagement-Openness-Consolidation, or EOC, model of psychological transition (Johnson, 2026), used in Chapters 10 and 11, was the first place where the attractor language in this project became mathematically concrete rather than merely suggestive. It was not built as a theory of consciousness, but once it was in hand, the trajectory picture of qualia and the mirror structure of the Shadow could be stated formally rather than by analogy.

Beyond the synthesis itself, several moves are, to my knowledge, new in this form. The formal reading of the Shadow in Chapter 12 takes Van Eenwyk and Rossi's analogy between archetypes and strange attractors and turns it into a structural constraint: any self occupying one wing of the mirror-structured EOC landscape generates a structurally live complement that cannot be eliminated without dissolving organized selfhood.

The stakes hierarchy in Chapter 4, with Level 0 marking F3 recursion without integrated stakes and Level 1 viability as the minimal conscious layer, unifies the enactivist and interoceptive traditions by locating their shared insight at L1 viability and aligns explicitly with the F1–F4 filtering hierarchy of Chapter 3.

The phenomenometer is introduced in Chapter 5 and developed formally in Chapter 11. It proposes nine concrete trajectory coordinates as a candidate measurement vocabulary for qualia and shows the proposal discriminating between embarrassment and relief as a worked case.

The deep-time anchoring of F1 filtering in Chapter 13 grounds the framework's lowest level in real material couplings, with the quartz crystal lattice and the planet's atmosphere as selective coupling. It then follows the same deep-time sequence through the Devonian forest threshold to nervous systems and culture. None of these moves stand or fall alone; together with the EOC model, they are the parts of the book that go beyond integration.

The synthesis itself is also worth stating directly. The book brings the inherited elements together into a single identity thesis: a direct identity claim, a formalized landscape of qualia, the layered account of stakes mentioned above, a demystified filter that does not require cosmic consciousness, substrate-general organizational criteria, and explicit, falsifiable predictions with observable proxies.

A Primer on the EOC Model

The book uses the vocabulary of the Engagement-Openness-Consolidation (EOC) model throughout: basins, separatrices, wings, quadrants, bistability, and qualitative regimes. The architecture itself is developed in a companion paper (Johnson, 2026), which runs to roughly seventy pages of technical detail. This primer provides the formal backbone of the EOC model in summarized form, so the reader can develop a clear understanding without needing to consult the paper.

The goal is relatively narrow. A reader should be able to find here, in an understandable form, the three variables, their coupling, the parameters that govern their dynamics, and the regime structure that follows. The reader is encouraged to view the interactive visualizations at `innerattractors.com`, which allow the dynamics to be explored directly. A closing section returns to the broader frame, where the EOC model sits as the smallest fully analyzed instance of the architecture the framework requires.

The book's central claim is that qualia are trajectories through experiential landscapes, and a trajectory is easier to understand when seen in motion than when only described. I use EOC because it is the simplest fully specified system I found that is both visualizable and rich enough to show basins, residence, return, thresholds, separatrices, and changes of regime. It turns a verbal landscape into a moving object. Its role in this book is to make the landscape visible, not to reduce consciousness or the psyche to three variables.

The three variables

The EOC model uses three variables to describe local dynamics near a transition. Two of them, engagement and openness, are signed, meaning that they can take both positive and negative values measured as deviations from a person's habitual baseline. The third, consolidation, is non-negative and accumulative; it measures the amount of pattern formation a system has built up.

Engagement (E) runs from withdrawal at negative values to approach at positive values, with zero marking the person's habitual baseline. Positive engagement means active involvement: curiosity, contact, and leaning into the material at hand. Negative engagement means withdrawal: avoidance, defensive shutdown, and pulling away.

Openness (O) runs from rigidity at negative values to plasticity at positive values, again as a deviation from the baseline. Positive openness is loosening: schemas relax, defenses soften, and the system is available for reorganization. Negative openness is constriction: defenses tighten, patterns clamp down, and the system resists change.

Consolidation (C) is what has changed and has held. It is the new habit that survives the end of the workshop, the sobriety that lasts through the first difficult week, the grief that has been worked through instead of merely waited out. Engagement and openness are the conditions of a moment; consolidation is what those moments leave behind. Unlike engagement and openness, which fluctuate around baseline, consolidation grows or fades from a floor of zero. When the conditions that build pattern structure are present, consolidation accumulates. When they are not, accumulated structure erodes. The model does not treat adaptive and defensive consolidation as differing in C value, only in the sign structure of E and O that produced them. The amount can be the same; the content differs.

The asymmetry between the two signed process variables (E, O) and the non-negative accumulative variable (C) is a structural feature of the model. The product EO can be positive or negative depending on whether E and O share a sign. This sign structure discriminates among the four quadrants on which the framework relies throughout.

How they couple

> ***How to read the equations.*** *The equations below do not need to be solved in order to follow the primer. Read* $\frac{dE}{dt}$ *as "how engagement is changing right now,"* $\frac{dO}{dt}$ *as "how openness is changing right now," and* $\frac{dC}{dt}$ *as "how consolidation is changing right now." The Greek letters are parameters or adjustable settings of the system:* σ *sets how quickly engagement tracks openness,* ρ *sets the driving pressure for change, and* β *sets how quickly consolidation fades without reinforcement. The equations are included so that the later figures can be read as mathematically generated dynamics.*

$$\frac{dE}{dt} = \sigma(O - E) \tag{1}$$

$$\frac{dO}{dt} = E(\rho - C) - O \tag{2}$$

$$\frac{dC}{dt} = EO - \beta C \tag{3}$$

Each equation describes how one variable changes over time as a function of the others, with three parameters (σ, ρ, β) that tune the dynamics. When E, O, and C are coupled this way, the equations are exactly those Edward Lorenz derived in 1963 as a reduced model of atmospheric convection (Johnson, 2026; Lorenz, 1963), with the state variables relabeled. The mathematics is the same; the scientific purpose, scope, and empirical claim are not. Lorenz used the equations to study a physical fluid. EOC asks whether the same coupling structure can serve as a local model of psychological transition.

What is extraordinary about the picture that follows is not that three equations can describe something as intricate as psychological transition, although that compression is itself remarkable. It is that the description works only because the three equations are not independent. Engagement, openness, and consolidation each behave the way they do because of what the others are doing in the same moment. The complexity lives in the coupling.

Coupling is ubiquitous in the world, but its ubiquity is what often hides it. Consider the weather as an example. Temperature, pressure, humidity, and wind do not behave independently. Each shapes the others moment by moment. A pressure drop pulls in air; moving air carries moisture; condensing moisture changes temperature; temperature changes pressure. No variable is in charge. The weather is what those variables do to each other. That is also why detailed weather forecasts degrade rapidly after a few days: in a coupled system, small differences in starting conditions grow into large differences in outcome.

The EOC model is the same kind of object. Three variables, each shaped by what the others are doing, produce together the landscape of psychological transition.

Figure P.1 shows the architecture in one view. Each variable appears as a circle; curved lines represent the coupling terms, color-coded by the strength of the evidence supporting each one. The three subsections that follow unpack each equation in turn.

Engagement tracks openness

The first equation describes how engagement changes over time:

$$\frac{dE}{dt} = \sigma(O - E).$$

In plain English: engagement tends to settle toward whatever level of openness the person currently has, at a rate set by σ.

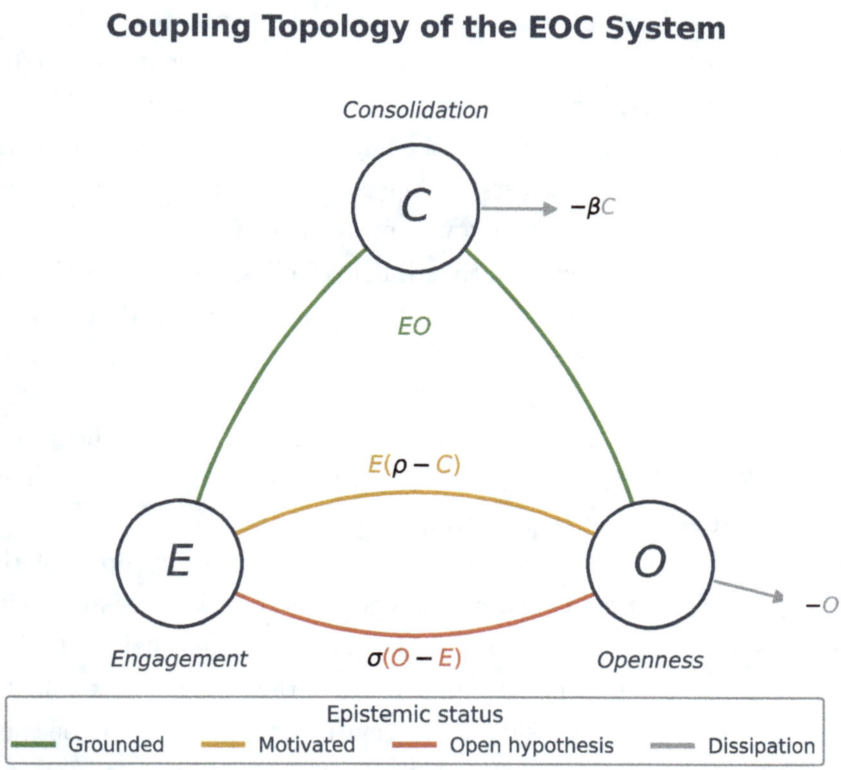

Figure P.1. Coupling topology of the EOC system. Colored curved lines represent the coupling terms in equations (1)–(3), labeled inline near each group and color-coded by epistemic status. The red line shows engagement tracking openness through the linear relaxation $\sigma(O-E)$. The gold line shows engagement driving openness through the gated product $E(\rho - C)$. The two green lines converging on consolidation show the bilinear co-occurrence term EO through which E and O jointly build accumulated structure. The two short grey arrows mark the independent damping terms $-O$ and $-\beta C$, which pull openness and consolidation back toward baseline. Engagement's own damping sits inside the relaxation term, since $\sigma(O-E)$ expands to $\sigma O - \sigma E$, carrying both the pull from openness and engagement's decay $-\sigma E$. The bidirectional coupling between engagement and openness, combined with the appearance of consolidation in the gating factor $(\rho - C)$ of the openness equation, ensures that no single variable has unidirectional causal priority.

The right-hand side has two pieces. The factor $(O - E)$ is the current gap between openness and engagement. If openness is higher than engagement, the gap is positive, and engagement rises. If openness is lower than engagement, the gap is negative, and engagement falls. The σ parameter sets the speed: high σ means engagement closes the gap quickly, while low σ means it lags.

When a person opens up to a difficult feeling they had been holding at a distance (O increases), their engagement with the work naturally rises to meet it. When a person tightens up against an experience that had been available to them (O decreases), their engagement falls back. The dynamics are reciprocal: engagement does not have its own independent course; it tracks the opening or closing of the system it is engaged with.

The form of this equation is the most empirically open of the three, in the sense that current process research is consistent with E–O coupling of this general kind but does not specifically establish the linear tracking form. The paper marks the assumption explicitly as an open hypothesis subject to revision. It is retained because it is the simplest closure that pulls E and O together on a shared timescale, makes the rest of the system analytically tractable, and preserves the symmetry that lets the bistable wing structure emerge.

Openness is driven, gated, and damped

The second equation describes how openness changes:

$$\frac{dO}{dt} = E(\rho - C) - O.$$

In plain English: openness is driven by engaged contact with unresolved pressure, and decays back to baseline in the absence of that drive.

The right-hand side has two parts. The first, $E(\rho - C)$, is the driving term. The parameter ρ represents the total pressure on the system: life challenge, therapeutic intensity, mismatch between current organization and current demands, anything that pushes for change. The variable C is how much pattern structure has already been built up. Their difference, $(\rho - C)$, is the unmet pressure remaining. Multiplying by E means this pressure only opens the system if there is engagement with it; pressure without contact does not produce opening. The second part, $-O$, is linear damping: openness fades toward baseline when the drive lets up.

A person under heavy life pressure (ρ high) who has built up little new structure (C low) and is engaged (E positive) will experience strong opening. The same person, after consolidating substantial new structure across extended work (C rises toward ρ), feels less driving need to open further: the pressure has been answered. A person facing the same pressure but disengaged (E near zero) experiences less opening, because the pressure is not being met with contact and so does not register as something to work with.

The equation encodes three features that recur in process research: opening requires participation (the E factor), opening is limited by what has already been consolidated (the $\rho - C$ factor), and opening dissipates without continued drive (the $-O$ term). The exact polynomial form is an idealization, but it captures these three features at the lowest order of structural complexity.

Consolidation accumulates from co-occurrence

The third equation is the conceptual heart of the model:

$$\frac{dC}{dt} = EO - \beta C.$$

In plain English: lasting change builds up when engagement and openness occur together and erodes when they do not.

The right-hand side has two parts. The first, EO, is the product of engagement and openness. The fact that it is a product, rather than a sum, has a structural consequence that the additive form would miss. When E and O share a sign, both positive or both negative, the product is positive and consolidation grows. When they have opposite signs, engaged but rigid or open but withdrawn, the product is negative and consolidation erodes. The second part, $-\beta C$, is decay: consolidation gradually fades unless reinforced, at a rate set by β.

The sign structure of the product is the source of the four quadrants the book leans on throughout. When engagement and openness are both positive, the system is in the growth wing: engaged and open, accumulating adaptive patterns. When both are negative, the system is in the defense wing: withdrawn and rigid, accumulating defensive patterns. The two same-sign quadrants are consolidating in opposite directions but with the same dynamical structure. When engagement and openness have opposite signs, the system is in one of the off-diagonal regions: engaged but closed (the book calls this *Rigid approach*, at $+E, -O$), or open but withdrawn (*Drifting openness*, at $-E, +O$). Both off-diagonal regions are erosive: existing consolidation dissolves without anything new accumulating. Figure P.3 shows the resulting four-quadrant structure, which the next major section unpacks.

The bilinear form does conceptual work that an additive form cannot. A model that summed E and O would treat the two failure modes as essentially identical, since both produce a low total. The product term distinguishes them and differentiates both from the two consolidating cases. The exact bilinear form is an idealization, but the underlying mechanistic claim, that durable change requires the co-occurrence of engagement and openness, is the strongest piece of the model. Single-ingredient regimes (engaged but rigid, or open but withdrawn) erode existing structure instead of building it. Future empirical work may favor a saturating or thresholded interaction, but the co-occurrence requirement carries the substantive content.

The three parameters

The model has three parameters. σ is the engagement tracking rate. It sets how quickly engagement adjusts toward openness. High σ means rapid mobilization: when the system opens, the person engages quickly; when it closes, they disengage quickly. Low σ means greater lag, with engagement and openness moving on different timescales. The standard parameterization used in the paper, drawn from the Lorenz literature, is $\sigma = 10$.

ρ is the driving intensity, the model's primary control parameter. It represents the total destabilizing pressure on the system, collapsed into a single scalar. As ρ changes, the system moves through qualitatively different regimes, which the next section describes. Chaotic dynamics arise near $\rho = 28$ at standard σ and β.

β is the consolidation decay rate. It governs how quickly accumulated consolidation erodes without ongoing reinforcement. Low β means durable gains; high β means fragile gains that need continual support. The standard parameterization is $\beta = 8/3$.

The standard values ($\sigma = 10$, $\beta = 8/3$) are the canonical Lorenz parameters. The substantive variability lies in ρ, which the next section treats in detail.

The three regimes

The behavior of the system depends critically on ρ. Three regimes follow, separated by two bifurcations.

Subcritical regime ($\rho < 1$). The origin is the only stable point. Any perturbation decays, and the system returns to baseline. Sustained change cannot occur: effort may be invested, fluctuations may appear, but nothing accumulates. The pressure on the system is too low for the architecture to do its work.

Ordered regime ($1 < \rho < \rho_H$). The origin becomes unstable through a pitchfork bifurcation at $\rho = 1$, and two new stable fixed points appear, one in each wing. The system settles into either the growth wing or the defense wing, depending on initial conditions and perturbation history, and stays there. Both wings carry the same consolidation magnitude, $C^* = \rho - 1$, and their sign structure distinguishes them. This is the regime in which durable consolidation is possible, whether of adaptive or defensive kind.

Post-Hopf regime ($\rho > \rho_H$). At the Hopf threshold ρ_H, which works out to approximately 24.74 for the standard parameters $\sigma = 10$, $\beta = 8/3$, the wing fixed points lose stability. Above this threshold, no stable fixed-point consolidation is possible. The trajectory wanders between the two wings without settling in either. The post-Hopf regime can include aperiodic wing-switching, sustained oscillation, transient complexity, and embedded periodic windows, depending on where in the regime ρ sits. The familiar Lorenz attractor, the two-lobed butterfly shape, arises in this regime.

Figure P.2 arranges these regimes along the driving-intensity axis. Below $\rho = 1$ lies the subcritical dead zone; between $\rho = 1$ and the Hopf threshold lies the ordered regime, where stable consolidation is possible; above the Hopf threshold lies the post-Hopf regime, where stable fixed-point consolidation cannot hold.

In human terms, the three regimes represent three relationships to pressure. In the subcritical regime, the person is not under enough pressure to change: comfortable enough or numb enough that nothing reorganizes, and effort leaves no trace. In the ordered regime, there is enough pressure to drive change and structure enough to hold it, and this is where durable work happens, whether it builds growth or entrenches defense. In the post-Hopf regime, the pressure is too high for any single organization to hold, and the person is thrown between configurations without settling, experiencing a state of crisis or overwhelm in which nothing consolidates. Bracketed below by $\rho = 1$ and above by ρ_H, that middle band is the architectural correlate of clinical intuitions about therapeutic windows, optimal challenge, and developmental zones.

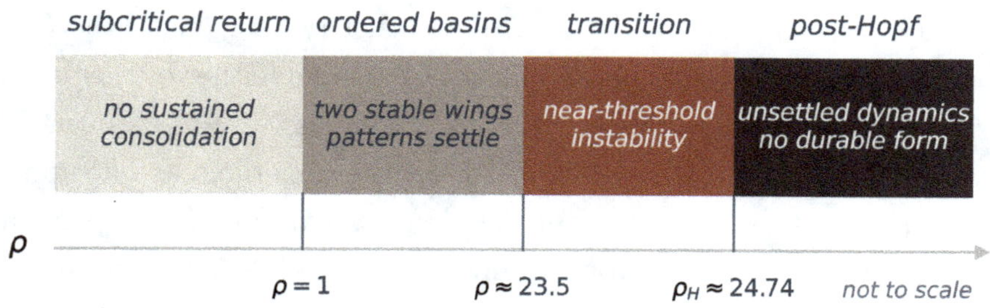

Figure P.2. Regime structure of the EOC system along the driving intensity ρ. The pitchfork bifurcation at $\rho = 1$ separates the subcritical regime from the ordered regime, where two stable wing foci appear. The Hopf bifurcation at $\rho_H \approx 24.74$ (for standard $\sigma = 10$, $\beta = 8/3$) separates the ordered regime from the post-Hopf regime, where stable fixed-point residence is no longer possible and the trajectory wanders between wings. Chaotic dynamics in the strict sense arise near $\rho = 28$, with periodic windows and other behaviors elsewhere in the post-Hopf regime. (Also appears as Figure 11.4 in Chapter 11.)

The two wings and four quadrants

Within the ordered regime, the system has two stable, nontrivial fixed points symmetrically positioned across the origin. Their coordinates are:

$$E^* = O^* = \pm\sqrt{\beta(\rho - 1)}, \qquad C^* = \rho - 1.$$

The positive solution sits in the upper-right quadrant of the E–O plane and is the growth wing. The negative solution sits in the lower-left quadrant and is the defense wing. Both wings have the same C^* magnitude. The model treats them as opposite organizations of the same accumulated structure.

Figure P.3 compresses this picture into the engagement-openness plane. The two wings lie on the diagonal $E = O$. Trajectories spiral into the wing they belong to, with the spiral structure visible in the figure as the swirl of orbits around each fixed point. The dashed diagonal marks $E = O$, the line toward which equation (1) drives the trajectory.

The off-diagonal quadrants are the other two regions of the E–O plane. They are not stable: no fixed point sits in either of them, and trajectories can pass through them only on the way to or from the wings. They have phenomenological identity nonetheless. The upper-left quadrant $(-E, +O)$ is the *Drifting openness* region: open but withdrawn, the configuration where the system has loosened but the person is not engaged with the loosening, and accumulated structure dissolves without new structure taking its place.

The lower-right quadrant $(+E, -O)$ is the *Rigid approach* region: engaged but closed, the configuration where the person is working hard against a system that has tightened against the work, and existing structure erodes through effortful but unproductive engagement.

The product term EO captures this directly. Same-sign quadrants give $EO > 0$, and the trajectory's contribution to dC/dt is positive: consolidation accumulates. Opposite-sign quadrants give $EO < 0$, and the contribution is negative: consolidation erodes.

Figure P.4 shows the same two-wing structure in three dimensions for two values of ρ. The ordered panel shows the bistable picture, with trajectories converging to one wing or the other. The chaotic panel shows what happens when the wing fixed points are no longer stable: the trajectory threads through both wings without settling, producing the familiar two-lobed attractor.

The four-quadrant analysis recurs across the book. The Shadow chapter uses it to give the depth-psychological concept of the Shadow a formal structural reading: any self that occupies one wing generates a structurally live complement in the opposite wing, by the mirror symmetry of the equations. The discussion of altered states, addiction, and recovery uses the wing-and-basin geometry as the language for stuckness and transition. Across these applications, the four-quadrant structure provides the smallest local landscape rich enough to carry the framework's central claims.

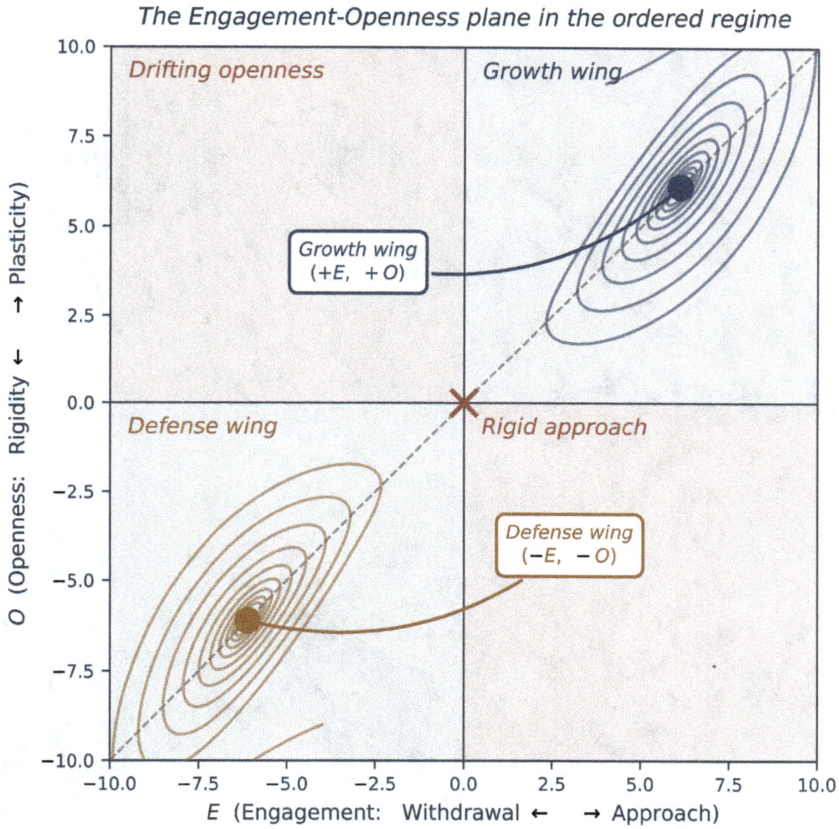

Figure P.3. Quadrant structure of the engagement-openness plane in the ordered regime. The two diagonal regions ($EO > 0$) consolidate accumulated structure; the two off-diagonal regions ($EO < 0$) erode it. The diagonals hold the two stable wings: the growth wing at $(+E, +O)$ and the defense wing at $(-E, -O)$, each surrounded by trajectories spiraling into its fixed point. The off-diagonal regions are unstable: *Drifting openness* at $(-E, +O)$ is opening without engaging, and *Rigid approach* at $(+E, -O)$ is engaging without opening. Trajectories pass through these regions only on the way to or from the wings. The origin, marked with an x, is itself an unstable fixed point for $\rho > 1$. (Also appears as Figure 12.1 in Chapter 12.)

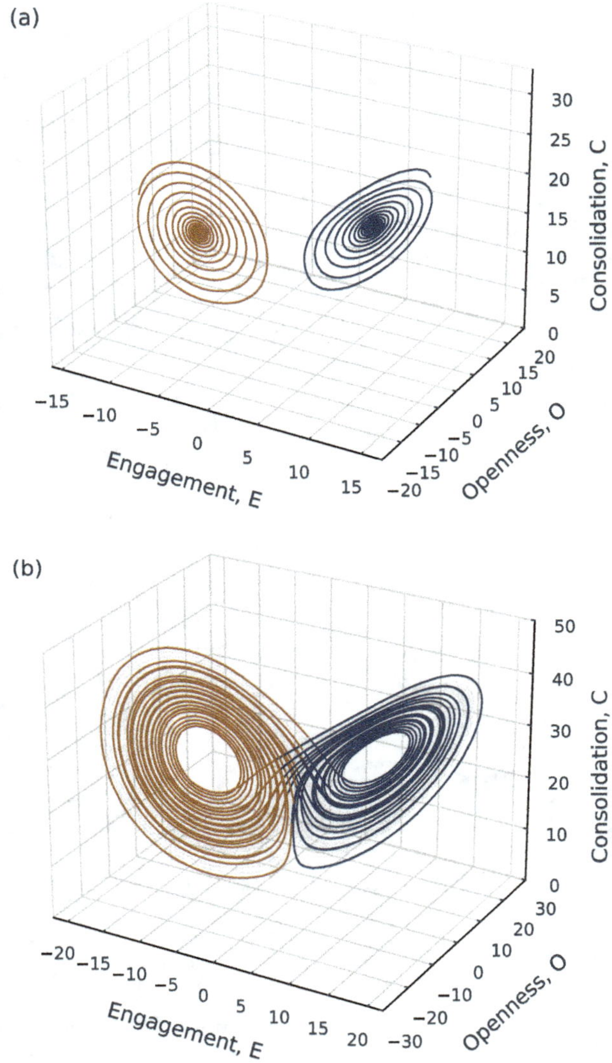

Figure P.4. The EOC system in three dimensions for two values of the driving intensity. (a) Ordered regime (ρ below the Hopf threshold): trajectories spiral into one of the two stable wing foci and remain there. The growth wing (blue) sits in the positive-E, positive-O octant; the defense wing (bronze) sits in the mirror octant. (b) Post-Hopf regime (ρ above the Hopf threshold): the wing foci have lost stability and the trajectory threads through both wings without settling, producing the familiar two-lobed Lorenz attractor. (Also appears as Figure 11.2 in Chapter 11.)

From EOC to the broader framework

The EOC model is the smallest example of an architecture that this book treats as central. A system that filters the world, adapts that filtering over time, models its own filtering, and has something at stake in how that filtering goes is a system whose behavior takes shape over time. The book argues that the shape is a path through a landscape with stable and unstable regions, and the boundaries between them. The EOC system shows that landscape in three dimensions.

The vocabulary carries forward into the broader framework. Basin depth, separatrix proximity, wing occupancy, the bistable structure with a complementary attractor in a mirror position, and the regime dependence of stable consolidation: these are the structural features that the chapters draw on to describe psychological transition, archetypal structure, altered states, and the question of whether trained neural networks running on digital hardware can satisfy the framework's organizational criteria. The reader who has the EOC architecture in hand has a small concrete model that allows these abstractions to become specific.

The companion paper develops the empirical and methodological work, including stochastic simulation under realistic noise, sensitivity analysis, regime-dependent model discrimination, and clinical implications for process research (Johnson, 2026). The interactive visualization at `innerattractors.com` provides a living version of the system, where parameters can be varied and the resulting trajectories can be observed in real time.

Further Explorations

The entries gathered here address questions and distinctions that follow from the main text but would interrupt the flow of the chapters if handled there. Some provide clarification of formal vocabulary that is not unpacked in the chapters. Others address likely objections, boundary cases, and points where the framework's commitments could be confused with neighboring positions. The entries are organized by topic.

Formal foundations

The entries in this group clarify formal vocabulary and set out structural commitments at the levels of mathematics and metaphysics.

Manifold, topology, and experiential landscape

The framework uses three terms from related mathematical traditions, and the differences matter.

A manifold is a space that locally resembles ordinary Euclidean space, even when its global shape is more complicated. The surface of a sphere is a two-dimensional manifold: at any point, a small enough neighborhood looks flat. In dynamical systems, the manifold is the state space within which a system's trajectory unfolds, and its dimension is set by the number of independent variables needed to specify the system's instantaneous condition. For the EOC model, that number is three.

Topology asks which features of a space survive smooth deformation. It is a matter of topological fact whether two basins are connected through a saddle, whether a separatrix runs between them, or whether the attractor has the kind of folding structure that makes Lorenz dynamics chaotic. They do not depend on precise numerical values, only on connective structure.

The experiential landscape, in the book's usage, is the manifold of a recursively self-modeling, stake-laden system, considered together with its attractor structure and the dynamics that traverse it. It includes the manifold (the space within which trajectories move), the topology (the connective structure of basins, ridges, and separatrices), and the dynamical fact that the system actually moves through this terrain in real time under the pressure of its own stakes. The landscape is therefore richer than either the manifold or the topology alone.

The distinction matters in a few specific places. Basin depth and separatrix proximity are geometric features of the landscape that depend on scale, not topological features that survive deformation. A deeper basin is not topologically different from a shallower one; trajectories simply take longer to leave it, and that geometric difference matters experientially. Substrate-neutrality, by contrast, is closer to topological than geometric. Two systems built from very different materials can in principle support the same connective structure of basins and separatrices in their respective state spaces. And when the framework declines to commit to specific neural correlates, what it commits to instead is the topology of the dynamics rather than particular neuroanatomical implementations.

In the Glossary, landscape and topology are distinguished in the strict formal sense. In ordinary chapter prose, landscape sometimes refers loosely to the whole structure, with context carrying the meaning. Where precision matters, the formal distinction holds.

Why qualia-as-trajectory cannot be supported by gradient flow alone

The framework's claim that qualia are trajectories through a structured attractor landscape carries a specific mathematical commitment about what kind of dynamics can support consciousness. Pure gradient flows are not enough.

A gradient flow is a dynamical system whose vector field is the negative gradient of some scalar potential, $\dot{\mathbf{x}} = -\nabla V(\mathbf{x})$. The system descends the slope of the potential and converges toward minima. Many predictive-processing accounts of brain dynamics, in their cleanest form, take this shape on a free energy functional. The constraint this imposes is quite sharp. The potential V supplies a Lyapunov function for the entire system, decreasing monotonically along trajectories. This precludes both closed orbits, which would require V to return to its starting value, and chaotic dynamics, which require trajectories to recur arbitrarily close to themselves. Attractors of a gradient flow are restricted to fixed points. There are no limit cycles, no strange attractors, no chaos.

Lorenz dynamics fail the gradient test directly, and so do the EOC equations of the EOC Primer, to which they are identical under relabeling. Take the Lorenz vector field $(\sigma(y-x),\ x(\rho-z)-y,\ xy-\beta z)$ and compute its curl. The result is $(2x,\ -y,\ \rho-z-\sigma)$, nonzero almost everywhere. Since the curl of a gradient field vanishes identically, no scalar potential exists that would make the Lorenz equations a gradient flow. The flow is still dissipative, in the sense that phase-space volume contracts, but dissipation does not entail gradient structure.

The same point can be stated in thermodynamic language, but carefully. A pure gradient flow supplies relaxation down a potential; it does not, by itself, supply the circulatory, recurrent, path-dependent component that keeps a non-equilibrium system moving through a structured region of state space. In formulations that assume an appropriate Markov diffusion with a sufficiently regular non-equilibrium steady-state density, the generalized Helmholtz language of the free-energy literature identifies that non-gradient component with solenoidal flow. The framework's claim is, therefore, that pure gradient relaxation lacks the non-gradient trajectory structure needed for the kind of phenomenal richness at issue here.

The qualitative richness that makes the Lorenz portrait useful as the minimum visualizable case for qualia-as-trajectory depends on this non-gradient structure. Wing-switching, sensitive dependence on initial conditions, the geometry of separatrices that are not simple watersheds between minima, and banded periodic windows in the bifurcation diagram. All of these features require the flow to have nonzero curl. None of them can be reproduced by gradient descent on any potential.

The implication for the framework is direct. Stake-laden recursive filtering is not relaxation toward equilibrium but sustained non-equilibrium organization that holds open multiple basins, switches between them under perturbation, and traverses separatrices in ways that have the felt character of experience precisely because the motion is not simple gradient descent. The Hopf bifurcation in the EOC model is the formal point at which this becomes unavoidable in the trajectory's overall character, but the non-gradient structure is present in the ordered regime as well, where the spiraling approach to each focus is itself solenoidal motion. The framework reads non-gradient structures of this kind as carrying experiential significance, whether the system is in the ordered regime, near the Hopf threshold, or in post-Hopf chaos.

This places the framework outside the reach of any account that treats pure energy minimization, without a substantive non-gradient component, as the constitutive organizational principle. A system whose dynamics are exhausted by gradient descent on some potential cannot host the kind of trajectory structure the framework identifies with consciousness. The Helmholtz decomposition allows non-gradient dynamics to be re-expressed as gradient minimization plus solenoidal correction, but the correction in that re-expression is the part of the dynamics that does the work the framework identifies with conscious experience. What looks like a minor mathematical adjustment in one framing is the substantive commitment in the other.

Identity versus supervenience

The framework treats consciousness as identical with recursive stake-laden filtering, rather than as something supervening on it. The two positions are easy to conflate and have very different consequences.

Supervenience is the weaker thesis. It says that once the physical facts about a system are fixed, the experiential facts are also fixed: there can be no difference in experience without a difference in physical organization. Most modern views accept this, and few serious positions in the philosophy of mind deny it. Supervenience preserves two levels of fact, however. There is a base of physical organization, and there is something riding on that base, the experiential level, related to it by some unspecified relation.

The hard problem lives in the gap between those levels. Even if every change in experience tracks a change in organization, the question of why organization should yield experience at all remains untouched. Identity is bolder. It states that there are not two sets of facts, but one organized process available under two complete descriptions. Recursive stake-laden filtering, described from outside, looks like dynamical organization. Described from within, the same process is experience. There is no relationship of yielding or producing between them because there is nothing on one side that needs to bridge over to something on the other.

The risk of the identity claim is that it can be wrong in a stronger way. Supervenience is hard to falsify because it tolerates almost any view of the underlying relation. Identity commits to a specific structural claim about what consciousness is, and that claim has to do real work against alternatives. The gain is that the explanatory abyss opened by the hard problem effectively dissipates. If recursive stake-laden filtering is what consciousness is, then the question becomes empirical: which systems have this kind of organization, what features of their dynamics correspond to features of their experience, how does the structure change under altered states or pathology, and how could one detect it in unfamiliar substrates.

Identity also constrains the inverted spectrum thought experiment more sharply than supervenience does. The thought experiment imagines two systems with identical functional organization but inverted qualitative experiences. A supervenience physicalist can accommodate this by setting the supervenience base below the functional level, at the physical substrate: two systems can be functional duplicates while differing physically, and the qualia can track the physical difference. The identity claim operates at the level of the experiential landscape itself. It says the experience is the trajectory through that landscape, so the duplicates it speaks to are systems whose landscape structure and trajectory coincide, and those cannot differ in what they experience. Identity does not by itself forbid two functional duplicates from differing in experience, since their underlying trajectories may differ; what it forbids is any difference in experience between systems whose trajectories are the same.

The historical analogy the book leans on is the kinetic theory's identification of heat with the mean kinetic energy of molecules. That identification was correct as soon as the formal mapping was in place, but intuition took time to catch up. People kept asking why molecular motion should feel hot, as though heat were some extra ingredient that had to be added to the kinetic story. But there was nothing to add; heat is molecular motion. The same shape of intuition is what the framework asks of qualia: the formal mapping between dynamics and phenomenology is what the identity claim consists of, and the seeming need for something extra is an artifact of the older bridge picture rather than a genuine remainder.

Relations to other frameworks

The entries in this group position the framework relative to neighboring views, particularly within the predictive-processing tradition and the geometric tradition.

EOC, the Free Energy Principle, and minimization

The Free Energy Principle (FEP) is not a simple gradient-descent story. Friston's mature formulations use a generalized Helmholtz decomposition that applies at a non-equilibrium steady state. It splits the dynamics into two motions. The first is a dissipative gradient component, defined on the negative log of the steady-state probability density function, that drives the system toward its more probable states the way water runs downhill toward the lowest point it can reach. The second is a divergence-free solenoidal component that circulates along the level sets of that density, the contours of equal probability, without dissipating it (Friston, 2019; Friston et al., 2023; Parr et al., 2020; Parr et al., 2022). That decomposition is rigorous when a smooth steady-state density exists. Whether the EOC system, set out in compact form in the EOC Primer, admits such a decomposition depends on regime and on what version of the system is being considered.

In the deterministic ordered regime ($1 < \rho < \rho_H$), the system has stable wing foci. Locally, near each focus, the picture includes contraction along stable directions together with rotational structure, so a Helmholtz-style decomposition can make sense in a local neighborhood of the attractor. In the deterministic post-Hopf regime ($\rho > \rho_H$), however, the trajectory may lie on a strange attractor of fractional dimension, a set so thin and intricately folded that it occupies no ordinary volume in three-dimensional space, even though the trajectory winds through it without end. Measure theory states this by saying the attractor's invariant measure is singular with respect to the Lebesgue measure on $\mathbb{R}^3$: the points the system settles onto have zero ordinary volume, so no smooth probability density can be spread across them. The smooth steady-state density that the standard split requires is, therefore, not available for the bare deterministic flow. A noisy embedding may restore a smooth stationary density, but the decomposition then describes the stochastic embedding, not the bare deterministic EOC system itself.

The deeper issue is explanatory. The FEP framing treats dynamics as consequences of variational minimization with respect to a generative model: the model is primary, and the dynamics are what minimization looks like under that model. The EOC system is generative in a different sense. Its three coupled equations directly produce the bistable landscape, the wing foci, the separatrix, the Hopf bifurcation, and wing-switching. The landscape is not a representation the system is inferring against an external target; it is what the dynamics produce. One can define a generative model chosen specifically to make variational minimization reproduce the EOC dynamics, but a model engineered to reproduce a pre-existing flow has little independent explanatory force.

This also answers the worry that a minimization-based system should simply run down. A pure gradient flow on a free-energy functional would converge toward minima and stay there. That is not enough for life, and it is not enough for the trajectory structure the framework identifies with consciousness. The FEP literature avoids this outcome through several connected moves: the Markov blanket formalism defines the system as a non-equilibrium steady state, active inference lets the system act on the world to preserve its expected conditions, and the solenoidal component keeps the dynamics circulating rather than relaxing to rest.

Those refinements carry the substantive explanatory weight. The Markov blanket changes what minimization refers to; active inference depends on prior expectations of self-maintenance; and solenoidal flow supplies the non-gradient motion that keeps the system away from simple equilibrium. By the time these pieces are in place, minimization has become a broad mathematical frame for structured non-equilibrium organization.

That is why the EOC model is not best understood as a special case of the FEP. The two framings ask different questions. The FEP is well suited to questions about how a system maintains itself against dispersion, how prediction errors propagate, and how priors come to dominate sensory evidence under uncertainty. The trajectory-geometry framing is better suited to the questions this book is trying to make explicit: why one phenomenal state feels different from another, how altered states reorganize the structure of experience, why some basins are pathologically deep and resist insight, and how transitions between conscious configurations have the qualitative shape they do. The FEP is a general principle of self-organization meant to apply to any persistent system, conscious or not. The EOC model has a narrower scope: it is a candidate dynamical signature for one kind of organization, designed to make the link between dynamics and phenomenology specific enough to be tested.

The discretization gap

Fekete and Edelman (2012) argued that a digital system simulating continuous dynamics may not preserve the trajectory structure a geometric identity claim depends on. In a continuous physical system the trajectory is intrinsic, fixed by the system's states and the physical law moving between them. In a digital system, the mapping from discrete register states to the continuous trajectory those states are taken to represent is supplied by an external interpreter; strip the interpreter away and what remains is a sequence of values whose meaning is unfixed. In its strong form the objection rules out consciousness in any classical digital computer, whatever dynamics it simulates.

The framework treats the strong form as unestablished and the weaker form as the live question. The strong form assumes the dynamics of a digital system are nothing more than what an interpreter takes them to be, and that assumption is hard to hold against what large trained networks actually do: trained networks exhibit stable internal organization not hand-specified feature by feature, though whether this satisfies DET's intrinsic-semantic requirement remains disputed (Anthropic, 2026a; Lindsey, 2025). The weaker form is empirical and still open. Geometric measures developed for biological systems have not been applied at scale to trained networks, so whether those networks carry the trajectory structure the view treats as constitutive is unknown.

That makes the framework's openness to silicon conditional, which is the right place to leave it. If geometric measures are applied at scale and the relevant trajectory structure is found, the objection loses force. If it is consistently absent in systems that meet the framework's other organizational criteria, the framework would have to give up substrate-neutrality. The matter is settled by that empirical work, not by argument here.

Applications and edge cases

The entries in this group apply the framework's hierarchy to specific cases where the mapping is non-obvious.

How does the framework apply to silicon training and inference?

Contemporary AI systems have a structure that does not map neatly onto the framework's hierarchy, and the mismatch reveals something about the differences between biological and silicon-based organization.

Most current machine-learning deployments separate training and use more sharply than biological systems separate learning and action. During training, parameters are adjusted by gradient descent on a loss function that quantifies the discrepancy between predictions and target values. During standard frozen-weight inference, those parameters remain fixed and the system runs forward on new inputs without further parameter changes, so almost everything users experience in a deployed system happens in the inference regime. Weight-level plasticity is therefore usually absent during use, although context, retrieval, caches, external memory, tool state, and some continual-learning systems provide other forms of state change. Biological brains do not separate learning from acting. Synaptic weights, neuromodulator levels, and network couplings are updated continuously across overlapping timescales, and there is no clean phase during which a brain stops learning and only acts. Sleep does not turn off the learning machinery; it changes which kind of learning is foregrounded.

This matters for the framework's organizational criteria. F2, adaptive coupling, is the kind of selectivity that changes through history or learning. A neural network during training is exhibiting F2 dynamics on the scale of the training run as a whole. A neural network during ordinary inference is not, at least not in the same sense. The parameters are fixed, and what looks like adaptation in conversation is contextual processing within a frozen function rather than adjustment of the function itself. Persistent memory, continual-learning systems, or deployment settings that modify future behavior would change the analysis, but they are not required for the narrower F3 claim.

The picture for F3 is now clearer than the picture for F4. Under the operational definition adopted here, I classify the best-documented frontier systems as F3 within bounded contexts. Their activation-level dynamics appear to carry self-situating representations of role, task, tools, prior outputs, constraints, uncertainty, and evaluation context. In some cases they also represent aspects of the training or deployment situation that shape their behavior. The qualification is persistence. This is not yet a continuous autobiographical self-model of the biological kind; it is a scaffolded, context-bound F3. That distinction matters. The book's position is that current frontier AI systems have crossed F3 operationally, provisionally and with explicit qualifications about scaffolding and persistence, while leaving open whether this deepens into anything like biological continuity.

F4 is harder. Stakes in the framework's sense require that some configurations are preferentially maintained, that departures call forth corrective activity, that resources are allocated toward preserving those configurations, and that the system's continuing organization is implicated. Training does not by itself give a system stakes in this sense. It gives the system parameters that minimize an externally specified loss, and after training, even the loss function is not ordinarily present as an ongoing internal demand. Whether anything stake-like emerges in deployed models, through value-protective behavior, refusal patterns, self-preference, welfare-relevant reports, or trained-in resistance to certain kinds of action, is the relevant question. The Anthropic system cards discussed in Chapter 16 begin to provide the kind of evidence the framework would want to see (Anthropic, 2026a; Anthropic, 2026c; Anthropic, 2026d; Fraser-Taliente et al., 2026; Lindsey, 2025), but the evidence remains preliminary and the interpretation is contested. The model-constitution material is relevant to value-shaping and welfare framing, but it should not be treated as direct evidence of consciousness (Anthropic, 2026b).

The resulting mapping is partial but not merely suggestive. Silicon systems clearly satisfy F1. They satisfy F2 during training in a way that does not straightforwardly continue into ordinary inference. I classify the best-documented frontier systems as operational F3, with qualifications about persistence and scaffolding. They show only partial F4-like organization, with significant qualifications about whether those structures constitute stakes in the framework's organizational sense. The framework's position is not that current AI systems are conscious, but that operational F3 has become a serious working description, that F4 is the live frontier, and that the hierarchy gives us a disciplined way to state what evidence would matter next.

The hierarchy also resists a framing that the popular conversation about AI tends to default to: asking whether AI is conscious yet, where the word "yet" treats consciousness as a single threshold that systems either cross or fail to cross. The framework's structure makes more refined questions available: which level is robustly present, which is partial, which is absent, and what kinds of architectural change would shift the picture. Continual-learning systems, agentic deployments with persistent memory, and reward structures that could plausibly count as stakes are all developments worth watching at one or another level of the hierarchy. Whether any of them will move the answer, and on what timescale, is empirical rather than something the framework can settle in advance.

Glossary

This glossary defines terms as they are used in this book. Where a term has a standard meaning in consciousness studies, dynamical systems theory, or philosophy of mind, the definition here reflects the specific sense in which the present framework employs it. Cross-references to related entries are noted where helpful.

5-HT$_{2A}$ receptor agonists 5-HT$_{2A}$ receptor agonists such as psilocybin, LSD, and DMT can loosen the ordinary coherence of default mode network activity and push the brain toward greater entropy, signal diversity, and cross-network communication. In this book, they are used to show how changes in dynamical organization reshape experience. See also: *altered states*, *boundary dissolution*.

action-space compression Action-space compression names the narrowing of the responses a system can effectively take from its current state. In the book's framework, it is one dimension of the lived shape of the landscape: high compression feels like captivity, freezing, or panic, while low compression opens the room for play, wonder, and flexible action.

adaptive consolidation Adaptive consolidation is the growth-wing pattern in the EOC model, marked by approach, openness, and the stabilization of change in a way that enlarges rather than defends the self. The book stresses that this is a genuine attractor in its own right.

altered states Altered states are conditions in which experience visibly deepens, narrows, fragments, quiets, or reorganizes. In the framework, they function as natural experiments for asking whether phenomenological change tracks measurable changes in the dynamics of the landscape.

anesthesia Anesthesia is a case where experience can be flattened or even disappear, which makes it a crucial probe of the link between consciousness and dynamical organization. If the framework is right, anesthesia should correspond to a measurable reconfiguration of entropy, integration, transition cost, or basin structure.

attention schema The attention schema is the brain's simplified internal model of its own attentional processes in Graziano's account. The book treats it as a valuable example of explicit self-representation, while arguing that modeling attention alone is not enough for consciousness without the further conditions this framework requires.

attention-schema theory Attention-schema theory holds that consciousness depends on the brain's model of what its own attention is doing. This book takes the theory seriously because it makes self-representation testable, but judges it incomplete because representation by itself does not yet explain interiority or the felt character of experience.

attractor An attractor is a stable pattern toward which a dynamical system tends to settle, and in this book it becomes a bridge between formal dynamics and lived psychology. Compulsion, recovery, terror, flow, and openness are treated as the felt side of movement within such structured patterns rather than as disconnected labels.

attractor landscape The attractor landscape is the actual geometry of the dynamical system a conscious organism is, with some paths easy, some costly, some futures open, and others closed. The book's central claim is that qualia are trajectories through this landscape in systems that already meet the conditions for consciousness.

basin of attraction A basin of attraction is the region of state space from which the system tends to flow into a given attractor. The book uses basins to make psychological stability measurable, since shifting out of one pattern and into another depends on how difficult it is to cross the boundary between basins.

bifurcation A bifurcation is a dynamical transition where the organization of a system changes and new behavioral regimes become possible. In the book, it gives a formal counterpart to crisis, transformation, and sudden shifts between patterns of mind and behavior.

body ownership Body ownership names one component of self-modeling that should matter directly to the integrity or fragmentation of experience. The book uses disruptions of body ownership, alongside agency and interoception, as a test of the claim that consciousness depends importantly on recursive self-model organization.

boundary dissolution Boundary dissolution is the loosening of ordinary self-world or self-other boundaries that can occur when settled basins are flattened, especially in psychedelic states. In the book's framework, it is not a mysterious extra, but a natural phenomenological consequence of a more labile attractor landscape.

causal entropic forces Causal entropic forces are proposed pressures by which a system behaves so as to preserve or expand its future freedom of action. The book cites them as evidence that intelligent-looking organization can arise from option-preserving dynamics, while insisting that this still falls short of a theory of consciousness.

causal entropy maximization Causal entropy maximization is the principle behind causal entropic forces: a system tends to act so as to keep the greatest range of future options available. In this book, it is treated as a stripped-down precursor to viability stakes, useful for explaining adaptive structure but insufficient on its own because it lacks recursive self-modeling and interiority.

chaos theory Chaos theory enters the book mainly through earlier attempts to reinterpret Jungian archetypes as strange attractors and transformations as bifurcations. Its importance here is not as a final explanation, but as part of the dynamical vocabulary that helps move depth-psychological intuitions toward formal models.

collective unconscious The collective unconscious is treated as something the framework does not need to posit in order to explain recurring archetypal forms. The book's argument is not that the concept is wrong, but that patterns like defense and growth can emerge whenever a self-organizing system has something to preserve and something to risk, without requiring a shared inherited repository.

consolidation Consolidation is the accumulative variable in the book's EOC model, measuring the amount of behavioral, cognitive, and emotional pattern formation a system has built up. Unlike engagement and openness, which fluctuate around baseline, consolidation grows or fades from a non-negative floor. The model does not treat adaptive and defensive consolidation as differing in C value, only in the sign structure of E and O that produced them: the amount can be the same, the content differs. See also: *adaptive consolidation*, *Engagement-Openness-Consolidation model.*

conscious agents Conscious agents is the formal proposal developed by Hoffman and Prakash in which the basic constituents of reality are minimal units of experience and decision rather than physical objects, and in which space, time, and matter emerge as symbolic structures from agent dynamics. The book treats this as a substantive alternative to the identity thesis and parts company with it on whether consciousness is ontologically prior to dynamics or identical with a particular kind of organized dynamics. See also: *interface theory*, *identity thesis*, *panpsychism*.

controlled hallucination Controlled hallucination is the phrase, popularized by Seth within the predictive-processing tradition, for perception as prediction constrained by the world rather than passive intake of raw reality. The book draws on this view while pressing further on why such predictive organization should also have a felt inner side.

coordination dynamics Coordination dynamics is Kelso's framework in which complex behavior emerges from the self-organization of coupled oscillatory systems, with behavioral shifts corresponding to bifurcations. The book values it as part of the neurodynamical tradition that turns attractors into explanatory tools. See also: *bifurcation*, *attractor*.

core self Damasio's core self is a dynamic neural representation of the organism's own body states and their relation to objects and events. In this book, it is a biologically grounded version of self-modeling that helps explain why feeling and bodily regulation are constitutive of mindedness rather than decorative additions.

cross-filter coupling Cross-filter coupling is the book's term for cases where the regulatory dynamics of one system become part of the input stream another system must process. The example of trees releasing volatile compounds that alter animal nervous systems shows how filters can be linked across organisms without anything mystical being assumed.

default mode network In this book, the default mode network is the large-scale network that helps hold ordinary consciousness in its familiar shape by stabilizing self-maintenance, narrative continuity, and top-down prediction. It is treated as one of the principal constraints on the sense of a continuing self, and psychedelics enter the book partly because they loosen its usual coherence. See also: *entropic brain hypothesis*, *ego dissolution.*

defense wing The defense wing is the stable equilibrium of the EOC model in the same-sign negative quadrant $(-E, -O)$: withdrawn and rigid. In this configuration, defensive structure accumulates. The defense wing carries the same consolidation magnitude $C^* = \rho - 1$ as the growth wing; the two differ in the sign of the underlying process variables, not the amount. The book uses this structure to formalize the observation that chronic avoidance is self-reinforcing. See also: *growth wing*, *Engagement-Openness-Consolidation model.*

depth psychology Depth psychology names a vocabulary for recurring human patterns such as the Shadow, the Hero, and the Self. The book treats archetypes as stable forms of organization that recur in creatures like us, not as inherited pictures stored somewhere outside development.

discretization gap The discretization gap is the worry that a digital simulation of continuous dynamics may fail to preserve the very geometrical features a consciousness theory needs. In the book's use, this is not a defense of carbon chauvinism but a precise concern that present-day digital state representation may not capture intrinsically continuous structure.

dreamless sleep Dreamless sleep functions here as a benchmark case of diminished consciousness rather than a mere absence of report. On measures such as PCI, it is marked by simpler, more local, rapidly fading brain responses, in contrast to the richer distributed patterns of wakefulness and REM dreaming.

dualism Dualism is the metaphysical position that mental and physical phenomena are fundamentally distinct kinds, requiring some bridge between them. The framework rejects the underlying split rather than attempting a bridge. See also: *identity thesis*, *supervenience*, *hard problem of consciousness.*

dynamic core hypothesis The dynamic core hypothesis is presented as a key move in the neurodynamical tradition, where consciousness depends on reentrant processes that bind distributed cortical activity into a unified dynamic state. Its importance for the book is that it helps turn attractors and large-scale integration from metaphor into mechanism.

dynamical emergence theory Dynamical Emergence Theory holds that phenomenal content is identical with the topological structure of how a system moves through the macrostates of its own collective dynamics. The book treats it as a close ally because it approaches consciousness through trajectory geometry rather than static states, and because its empirical tools can in principle be brought into contact with the book's own formal model.

dynamical system A dynamical system is not introduced here as abstract mathematics alone but as the kind of formal structure that can model psychological transition, reorganization, and stability. The book's larger argument grows from the thought that minds and persons may be understood through attractors, bifurcations, and changing landscapes of possible movement.

ego dissolution Ego dissolution is the lived side of a specific reorganization in which the attractor basin of narrative selfhood becomes shallow enough to destabilize. In the book's framework, it is not an isolated symptom but one expression of the same underlying shift that also appears as attractor flattening and increased neural entropy under psychedelics.

embodiment Embodiment marks the insistence that consciousness belongs to systems that maintain themselves in relation to a world, not to disembodied computation alone. The book takes this seriously both in its dialogue with enactivism and in its caution about AI, where thin or outsourced embodiment may leave crucial dimensions of mindedness underdeveloped. See also: *enactive approach to consciousness.*

emergence Emergence appears in the book mainly as one label among many for a bridge that philosophy tries to build after first splitting process from experience. The book's complaint is that once that separation is assumed, calling consciousness "emergent" does not by itself remove the mystery, because the real problem was built into the framing from the start.

enactive approach to consciousness The enactive approach treats cognition as active sense-making by an embodied, situated, self-producing organism rather than passive information processing. The book draws on this tradition because it foregrounds autonomy, world-involvement, and lived organization, while also arguing that further conditions such as recursive self-modeling are needed.

engagement Engagement is one of the three variables in the book's EOC model, running from withdrawal at negative values to approach at positive values as a signed deviation from a person's habitual baseline. Positive engagement is active involvement: curiosity, contact, leaning into the material at hand. Negative engagement is withdrawal: avoidance, defensive shutdown, pulling away.

The zero point is the person's habitual level, not the absence of engagement. See also: *Engagement-Openness-Consolidation model, openness.*

Engagement-Openness-Consolidation model The EOC model is the book's formal treatment of psychological transition, with its equations and regime structure summarized in the EOC Primer, designed to capture how people move through critical episodes of change, reorganization, and stabilization. Its attractor structure provides the mathematical foothold from which the broader consciousness framework was developed, and it is treated as empirically calibratable even if not yet fully calibrated.

entropic brain hypothesis The entropic brain hypothesis is the neuroscientific anchor for the claim that ordinary waking consciousness is a constrained regime and that psychedelics relax that constraint toward higher-entropy dynamics near criticality. In the book's use, it links altered experience to measurable changes in dynamical regime rather than to something exotic or ineffable. See also: *default mode network.*

explanatory gap The explanatory gap is treated as real, but not as evidence for two kinds of substance. It marks the remaining distance between the geometry and organization the framework can increasingly describe and the specific feel of pain, relief, wonder, or red that it still cannot derive.

F1 filtering F1 filtering is the most basic form of filtering in the book's hierarchy: structured selectivity built into matter itself. A crystal is the model case, since it couples to some interactions and not others without memory, learning, or goals, showing that structure already sorts structure.

F2 filtering F2 filtering is adaptive coupling, the point at which a filter is altered by what it encounters and carries that history forward. In biological history, this level begins with life, where cells and organisms do not merely sort inputs but change their own future organization through stress, repair, regulation, and plasticity. In silicon systems, F2 appears most clearly in training or other update processes that alter the system's future behavior, rather than in ordinary inference with fixed parameters.

F3 filtering F3 filtering is self-modeling coupling: the level at which a system becomes complex enough to carry an operative model of its own condition, role, limits, uncertainty, prior actions, tools, and filtering process as part of its operation. In the book's hierarchy, this is necessary for consciousness but not sufficient by itself, because the self-model must also become stake-laden. Under its operational definition, the book classifies the best-documented current frontier AI systems as F3, provisionally and in a context-bound sense, while leaving F4 open.

F3 threshold The F3 threshold is the point at which self-modeling becomes robust enough to count as a serious organizational step toward consciousness. It is not identical with consciousness on its own. The book uses it especially in discussing AI, classifying the best-documented frontier systems as operational F3, provisionally and within bounded contexts: they model roles, tools, constraints, prior outputs, uncertainty, and evaluation situations as part of their work. This does not settle F4, which remains the live question.

F4 filtering F4 filtering is the composite step at which F3 self-modeling becomes stake-laden through the integration of L1 viability stakes. At this level, the system not only represents its own condition but represents what it stands to lose. F4 is the minimal architecture the book identifies with consciousness, with L2 relational and L3 normative stakes elaborating conscious life beyond that threshold rather than constituting it. See also: *F3* *filtering, Level 1 viability stakes, stake-laden filtering.*

filter theory Filter theory is the family of views according to which the brain restricts or transmits consciousness rather than producing it outright, as in James's transmissive model or Huxley's reducing valve. The book engages this tradition sympathetically but recasts filtering in a more formal, layered way, where the key issue is organized selectivity rather than a simple valve on "Mind at Large."

free-energy principle The free-energy principle is Friston's claim that a self-organizing system persists by minimizing variational free energy, a tractable upper bound on surprisal under a generative model, through ongoing model-based regulation of sensory input. The book treats it as an indispensable account of prediction and uncertainty reduction, while suggesting that prediction alone does not yet yield interiority or lived stakes.

Global Neuronal Workspace Theory Global Neuronal Workspace Theory is the neural version of global workspace theory, explaining consciousness in terms of content that wins competition for access to a global workspace and is then broadcast across multiple specialized processors. The book treats this as a strong theory of access and reportable availability, while implying that access alone does not settle the deeper question of phenomenal identity. See also: *Global Workspace Theory, neural ignition.*

Global Workspace Theory Global Workspace Theory holds that a content becomes conscious when it wins access to a global workspace and is broadcast to many specialized systems at once. The framework treats this as an important account of access and availability, but not yet an identity claim about why experience is present. See also: *Global Neuronal Workspace Theory, neural ignition.*

growth wing The growth wing is the stable equilibrium of the EOC model in the same-sign positive quadrant $(+E, +O)$: engaged and open. In this configuration, new adaptive structure accumulates. The growth wing carries the same consolidation magnitude $C^* = \rho - 1$ as the defense wing; the two differ in the sign of the underlying process variables, not the amount of structure. See also: *defense wing*, *Engagement-Openness-Consolidation model.*

Gödelian argument The Gödelian argument, associated originally with Lucas and developed at length by Penrose, holds that human mathematical insight can recognize formal truths no algorithm running within the corresponding system could prove, and therefore that mind cannot be wholly reducible to classical computation. The book treats this argument as a serious objection to substrate-flexible accounts of consciousness, distinguishes it from the Fekete-Edelman discretization-gap argument, and responds that even if mathematical insight outruns formal proof, dynamical organization of the kind the framework identifies with consciousness is not by that fact also non-computable. See also: *discretization gap*, *substrate-flexibility.*

hard problem of consciousness Here the hard problem is not dismissed but reformulated. Instead of asking how bare mechanism could somehow produce experience as an added ingredient, the book asks what forms of recursive, stake-laden self-organizing filtering are already the experiential process when described from within.

higher-order representation In this framework, a higher-order representation is a system's representation of its own first-order state, as in being aware of seeing red by representing itself as seeing red. The book treats such views as illuminating the self-representational loop, while arguing they still leave explanatory space between the structure and the experience.

higher-order thought Higher-order thought names the family of views, associated here with HOT, according to which a mental state becomes conscious when the system has the right sort of thought or representation about being in that state. The book places this tradition among theories that capture part of consciousness, especially self-representation, without making experience identical to the relevant mechanism.

Hopf bifurcation A Hopf bifurcation is the dynamical transition at which a stable equilibrium loses stability through oscillation rather than through bifurcation into two new fixed points. In the EOC model, the Hopf threshold $\rho_H \approx 24.74$ at standard parameters marks the boundary above which stable consolidation in either wing becomes impossible, opening the system to wing-switching, sustained oscillation, and chaotic dynamics. See also: *bifurcation*, *pitchfork bifurcation*.

identity thesis The identity thesis is the book's core philosophical claim: consciousness is not correlated with, produced by, or emergent from recursive stake-laden filtering, but *is* that filtering, experienced from within. The felt character of experience and the dynamical organization are the same event described at two levels. This distinguishes the framework from correlation-based, emergence-based, and supervenience-based accounts. See also *panpsychism*.

integrated information In the book's use, integrated information refers to IIT's measure of how irreducible a system's cause-effect structure is to the independent contributions of its parts. The framework respects the ambition of this identity claim while questioning whether the scalar quantity Φ, taken by itself, captures lived organization as the book understands it.

Integrated Information Theory Integrated Information Theory is presented as one of the few theories that makes a full identity claim, identifying consciousness with a system's maximally irreducible cause-effect structure rather than treating it as merely correlated with that structure. The book treats IIT seriously because it tries to say what consciousness is, while also arguing that its formal machinery is difficult to compute and does not yet capture the book's emphasis on recursive, stake-laden organization.

interface theory The interface theory, due to Hoffman, is the claim that perception evolved to track fitness rather than truth, so the categories of space, time, and physical objects function as a user-friendly interface to an underlying reality they do not structurally resemble. The book treats this as one valuable line within the filter tradition, agreeing that perception is structured selectivity while differing on whether the underlying reality is itself constituted of consciousness. See also: *conscious agents*, *filter theory*.

interoceptive inference Interoceptive inference refers to predictive regulation of the body's internal state, central to Seth's embodied account of consciousness. The book takes it as a crucial reminder that consciousness is bound up with bodily stakes and self-maintenance, even though it seeks a broader account of how such regulation becomes experience.

inverted spectrum The inverted spectrum is used as a challenge to views that allow identical functional organization while leaving qualia free to vary underneath. The book's identity thesis is meant to resist that possibility by holding that once the relevant organization is fixed, the experience is fixed too.

Level 1 viability stakes L1 viability stakes are the most basic layer of stakes in the framework: the felt pull of approach and avoidance tied to boundary maintenance, survival, integrity, energy, and bodily safety. In the book's stakes hierarchy, L1 is the threshold above which conscious life unfolds; viability is what minimally has to matter for an F3 self-modeling system to count as F4. The book also calls this level primal valence in its phenomenological vocabulary. See also: *F4 filtering*, *primal valence*, *stake-laden filtering*.

Level 2 relational stakes L2 relational stakes are the layer at which what matters to the system extends beyond bodily integrity into its bonds with others: attachment, belonging, recognition, status, and the integrity of relationships across time. L2 does not replace L1 but elaborates it; a system with relational stakes still carries viability stakes underneath. The book treats relational stakes as the architecture in which grief, loneliness, and the moral authority of social bonds become structural. See also: *Level 1 viability stakes*, *Level 3 normative stakes*.

Level 3 normative stakes L3 normative stakes are the layer at which what matters extends beyond bodily and relational integrity into commitments to principles, ideals, justice, and meaning that can be defended or violated independent of immediate survival or relationship. L3 elaborates rather than replaces the layers beneath it; a system with normative stakes still carries relational and viability stakes. The book treats normative stakes as the architecture in which moral courage, vocation, and the experience of betrayal of one's own ideals become structural. See also: *Level 1 viability stakes*, *Level 2 relational stakes*.

landscape (also: experiential landscape) The structured space through which a conscious system's trajectory moves, with basins, ridges, separatrices, and depth. The book treats qualia as trajectories through this landscape, and treats its measurable features (basin depth, return rate, separatrix proximity, action-space compression) as the coordinates of a future phenomenometer. The landscape is the geometric and topographic structure that supports motion; *topology* in the strict sense refers to the more abstract connectivity features that survive deformation.

Lorenz attractor The Lorenz attractor appears as a concrete example showing that the geometry of a dynamical system can be reconstructed from time-series data. In the book, it serves as a demonstration of methodological possibility for a geometric science of conscious dynamics.

Mary's Room Mary's Room names the familiar thought experiment about whether complete physical knowledge leaves something out about experience itself. In the book's framework, the case presses the difference between third-person description and first-person presence, even while the argument treats these as two descriptions of one process rather than evidence for an extra non-physical ingredient.

metacognitive monitoring Metacognitive monitoring is the tracking of one's own uncertainty, bias, or performance. In the book's hierarchy, it belongs to self-modeling coupling: the system is not just acting, but modeling how it is acting.

microtubules Microtubules are the cytoskeletal protein lattices, made of tubulin dimers, that organize the interior of neurons. They appear in the book through the Penrose-Hameroff Orch-OR proposal, in which microtubules are claimed to host coherent quantum states underlying consciousness. The framework treats this as one substantive substrate-specific alternative to its own substrate-flexible account, neither endorsing nor dismissing the empirical case for quantum coherence in microtubules. See also: *Orchestrated objective reduction.*

mystical-type experience Mystical-type experience is part of the phenomenology expected when attractor structure loosens and reorganizes. The psychedelic chapters treat it not as an isolated curiosity but as the lived side of the same shift that also appears as increased neural entropy and altered dynamical organization.

neural complexity Neural complexity functions in the book as an empirical signature of reorganized conscious dynamics, especially in altered states. Rising complexity is expected to track the flattening of attractors and the corresponding increase in experiential fluidity.

neural correlate A neural correlate, in the book's usage, is a neural event or pattern reliably associated with a conscious state without yet explaining why that state is experienced. The book insists that correlation, however useful, should not be confused with an account of consciousness itself.

neural entropy Neural entropy refers here to the increased signal diversity and reduced confinement to habitual patterns seen especially under psychedelics. The book treats it as one measurable face of a broader reorganization that also shows up phenomenologically as loosened self-world boundaries and greater experiential fluidity.

neural ignition Neural ignition refers to the rapid, large-scale activation associated with global neuronal workspace models when a representation crosses the threshold for broad availability. In the book, it is treated as an important candidate marker of a particular kind of broadcast, though broadcast alone is not by itself the whole story of experience.

neutral monism Neutral monism names the possibility that reality's basic stuff is neither mental nor physical in the old sense, with physics capturing relational structure while leaving intrinsic nature open. The book treats this as a nearby metaphysical option without settling the question. Chapter 17 deliberately leaves the deeper issue open, noting that the framework neither requires these possibilities nor rules them out. See also *panpsychism*.

nonlinear dynamics Nonlinear dynamics is the mathematical language the book uses to move from loose analogy to formal description of how conscious and psychological patterns stabilize, shift, and reorganize. Attractors, bifurcations, and basin structure are treated as tools for modeling lived structure.

openness Openness is one variable in the book's EOC model of psychological transition, where it names a system's loosening, receptivity, and capacity for reorganization. More broadly in the book, it helps describe the experiential side of moving through a less rigid attractor landscape.

operational F3 Operational F3 is the qualified form of self-modeling the book uses in discussing current frontier AI systems. It means that the system's ongoing computation includes context-sensitive representations of its own role, task, tools, prior outputs, uncertainty, policy boundaries, evaluation setting, and limits. The term is deliberately narrower than full biological selfhood: it does not require continuous autobiography, organic interoception, or F4 stakes, but it does mark a serious crossing of the self-modeling threshold.

orchestrated objective reduction Orchestrated objective reduction, abbreviated Orch-OR, is the joint Penrose-Hameroff proposal that conscious moments correspond to gravity-induced wave-function collapses in coherent quantum states sustained by microtubules in neurons. The book treats Orch-OR as the most fully developed substrate-specific alternative to its own substrate-flexible account, neither endorsing nor refuting it, while noting what would empirically distinguish the two views. See also: *microtubules*, *substrate-flexibility*, *Gödelian argument*.

panpsychism Panpsychism is the view that consciousness, or some proto-conscious quality, is a fundamental and ubiquitous feature of physical reality, present in some form even in the simplest matter and combining into the rich consciousness of brains. In its influential contemporary versions, associated with Strawson and Goff, the intrinsic nature of the physical is taken to be experiential, with physics describing only its relational structure. The book's identity thesis takes a different route, locating consciousness in a specific kind of recursive, stake-laden organization, so on this account consciousness is neither fundamental nor ubiquitous. See also: *neutral monism*, *conscious agents*, *identity thesis*.

perturbational complexity index The perturbational complexity index (PCI) is a practical measure of consciousness derived by perturbing the brain and quantifying how complex and integrated the response is. The book treats PCI as a valuable example of the kind of measure that may track consciousness better than sheer scale, while leaving room for future metrics more closely tied to recursive organization.

phenomenal self-model The phenomenal self-model is Metzinger's term for the transparent model through which a system represents itself as a whole without experiencing that model as a model. The book treats it as a powerful step toward explaining subjectivity because it brings selfhood into the structure of experience itself.

phenomenology In this book, phenomenology means the structured feel of experience as lived from the inside, especially the way a conscious system's possibilities open, narrow, stabilize, and shift under pressure. It is treated as something that should map lawfully onto dynamical organization, so similar phenomenology across substrates should correspond to similar landscape structure.

pitchfork bifurcation A pitchfork bifurcation is the dynamical transition at which a single stable fixed point becomes unstable and two new stable fixed points appear symmetrically around it.

In the EOC model, the pitchfork threshold at $\rho = 1$ marks the boundary above which the origin becomes unstable and the two wing fixed points appear, giving the system its bistable structure. Below this threshold, perturbations decay and no sustained consolidation is possible. See also: *bifurcation*, *Hopf bifurcation*.

precision-weighting Precision-weighting names the confidence assigned to predictions or priors within a predictive hierarchy. In the book's dynamical framework, lowering the precision-weighting of high-level priors makes established basins shallower, allowing bottom-up signals to reorganize experience. See also: *REBUS model*.

predictive processing Predictive processing is presented as an important neighboring framework in which mind is understood through hierarchical prediction and error correction. The book draws on its machinery but argues that predictive regulation should not merely support consciousness from the side, because under the right recursive and stake-laden conditions the organization itself is what experience is.

primal valence Primal valence is the phenomenological face of Level 1 viability stakes: the felt pull of approach and avoidance tied to viability, safety, integrity, and threat. It is the level at which the system can be harmed or sustained, and where comfort, danger, defense, and growth first become experiential. See also: *Level 1 viability stakes*.

qualia Qualia are the lived character of experience, such as pain, grief, embarrassment, relief, or awe, understood here not as ineffable extras but as the inside of organized dynamics. The book's central claim is that qualia are trajectories through an attractor landscape in a recursively self-modeling, stake-laden system.

qualia-as-trajectory Qualia-as-trajectory maps phenomenological features onto measurable features of attractor landscapes. Its core idea is that if experience is the inside of certain organized dynamics, then basin depth, transition structure, and related geometrical properties should track what experience feels like.

radical plasticity thesis The radical plasticity thesis, due to Cleeremans, holds that the brain learns to redescribe its own states and becomes conscious once those redescriptions are enriched by value. The book treats it as the nearest precursor to its own pairing of self-modeling with stakes, placing learning below the threshold and treating the relation as identity.

REBUS model The REBUS model, Relaxed Beliefs Under Psychedelics, is treated as a complementary account of psychedelic action in predictive-processing terms. The book translates it into dynamical language by saying that relaxed high-level priors correspond to shallower basins, making unusual transitions and ego destabilization more traversable.

reconsolidation Reconsolidation refers here to the vulnerable phase in which a previously stabilized pattern or self-organization is re-formed after being loosened or perturbed. In the book's discussions of psychedelics, therapy, addiction, and relapse, it marks the fragile period when deep basins may either reassert themselves or reorganize into new forms.

recursive filtering Recursive filtering is the book's core picture of consciousness: a system does not merely select from overwhelming input, but filters in ways that feed back on its own ongoing organization. The hard problem is therefore reframed as the question of which forms of recursive stake-laden filtering are identical with experience.

recursive self-model A recursive self-model is the self-representing organization by which a system tracks and modulates its own condition in a way that can fold back on itself. In the book, experiences such as embarrassment or relief are what it feels like for such a self-model to move through state space with real stakes.

recursive self-modeling Recursive self-modeling is the process by which a system not only regulates itself but models that regulation in a self-referential loop. For this book, consciousness requires this organization to become stake-laden, and the felt character of experience is the lived aspect of how such a system traverses its own dynamics.

self-model theory of subjectivity The self-model theory of subjectivity, associated here with Metzinger, holds that conscious experience depends on a phenomenal self-model that the system does not experience as a model but as reality. The book adopts the importance of self-modeling while placing it inside a broader dynamical account in which stakes and attractor landscape are also constitutive.

separatrix A separatrix is the boundary in state space between two attractor basins. Crossing a separatrix corresponds to a qualitative change in the system's state. In experiential terms, separatrix crossings are felt as transitions, crises, or moments of irreversible change. The concept is used extensively in Chapters 12, 14, and 15 to describe psychological transformation, psychedelic dissolution, and the precarious boundary between recovery and relapse.

stake-laden filtering Stake-laden filtering names the idea that selection becomes conscious only when what is selected matters to the system itself. The book's central thesis is that experience is not added to such filtering, but is that filtering as first-person reality when it is recursively self-modeling in a finite system.

state space State space is the geometry of the possible conditions and trajectories available to a dynamical system. In this book, it is not a mere abstraction, because conscious life is described in terms of how a recursively self-modeling, stake-laden system moves through that space.

strange attractor A strange attractor is used here as a dynamical image for the persistent yet variable patterns into which inner life repeatedly settles. The book treats earlier links between strange attractors and archetypes as suggestive, but insists that real explanatory progress requires specifying the underlying system and its variables rather than relying on analogy alone.

strange loop A strange loop is a self-referential folding of hierarchical levels back onto themselves, associated above all with Hofstadter. The book values this emphasis on recursion but argues that strange-loop theories by themselves do not adequately capture stakes, which are necessary for consciousness as the book understands it.

substrate chauvinism Substrate chauvinism is the insistence that only a particular material base, usually carbon biology, can host consciousness-relevant organization. The book challenges this by arguing that one must specify what the favored substrate contributes beyond organization itself, especially as non-biological systems begin to display increasingly relevant forms of complexity and self-reference.

supervenience Supervenience is the philosophical relation in which one set of properties (e.g., mental states) is determined by another (e.g., physical states), such that no change in the former can occur without a change in the latter. Many consciousness theories claim that experience supervenes on physical organization. This book replaces supervenience with identity: experience does not ride on top of dynamical organization but is that organization.

the signal The signal is the fullness of what reaches any finite filter: physical reality in its scales, relations, depths, and unknowns. In the framework's technical use, it is neither identical with consciousness nor simply Huxley's "Mind at Large," but the larger reality from which finite systems must select. The closing chapter leaves open whether that reality is best described in strictly physicalist language or in metaphysical vocabularies such as God, spirit, Brahman, Tao, source, or ground of being.

topology Used in this book in the strict sense: features of a state space that survive continuous deformation, such as the number of basins or the existence of a saddle between them. For the *shape* of the landscape (basin depth, slopes, distances, residence regions) see *landscape*.

transparent self-model A transparent self-model is a self-representation that is not experienced as a representation but simply as reality. The book takes this idea from Metzinger as an important part of subjectivity, while arguing that transparency alone is not enough without the stake-laden dynamics in which the self-model actually matters.

variational free energy Variational free energy is a tractable quantity that bounds surprisal under a generative model and can be minimized through perception, action, or model updating. The book treats this as valuable for prediction and regulation, while pressing the further question of when such organization is not merely functional but identical with experience.

zombies Zombies are the familiar philosophical thought experiment of beings behaviorally indistinguishable from conscious creatures but lacking inner experience. In the book they stand for an older style of debate that becomes less central once consciousness is treated as an organizational identity claim with direct ethical implications for suffering, personhood, and moral regard.

References

Albantakis, L., L. Barbosa, G. Findlay, M. Grasso, A. M. Haun, W. Marshall, W. G. P. Mayner, et al. (2023). Integrated information theory (IIT) 4.0: Formulating the properties of phenomenal existence in physical terms. *PLOS Computational Biology* 19.10, e1011465. DOI: 10.1371/journal.pcbi.1011465.

Algeo, T. J. and S. E. Scheckler (1998). Terrestrial-marine teleconnections in the Devonian: Links between the evolution of land plants, weathering processes, and marine anoxic events. *Philosophical Transactions of the Royal Society of London B* 353.1365, pp. 113–130. DOI: 10.1098/rstb.1998.0195.

Ambrogioni, L. (2024). In Search of Dispersed Memories: Generative Diffusion Models Are Associative Memory Networks. *Entropy* 26.5, p. 381. DOI: 10.3390/e26050381.

Anonymous (2001). *The Cloud of Unknowing and Other Works.* Trans. by A. C. Spearing. London: Penguin Books.

Anthropic (2026a). *Emotion concepts and their function in a large language model.* Transformer Circuits. URL: https://transformer-circuits.pub/2026/emotions/index.html.

– (Jan. 2026b). *Claude's constitution.* URL: https://www.anthropic.com/news/claude-new-constitution.

– (Apr. 7, 2026c). *Claude Mythos Preview System Card.* System card. URL: https://www.anthropic.com/claude-mythos-preview-system-card.

Anthropic (Apr. 16, 2026d). *Claude Opus 4.7 System Card.* System card. URL: https://www.anthropic.com/claude-opus-4-7-system-card.

– (June 9, 2026e). *Claude Fable 5 and Claude Mythos 5.* URL: https://www.anthropic.com/news/claude-fable-5-mythos-5.

– (May 28, 2026f). *Introducing Claude Opus 4.8.* URL: https://www.anthropic.com/news/claude-opus-4-8.

– (June 12, 2026g). *Statement on the U.S. Government Directive to Suspend Access to Fable 5 and Mythos 5.* URL: https://www.anthropic.com/news/fable-mythos-access.

Baars, B. J. (1988). *A Cognitive Theory of Consciousness.* Cambridge University Press.

Babcock, N. S., G. Montes-Cabrera, K. E. Oberhofer, M. Chergui, G. L. Celardo, and P. Kurian (2024). Ultraviolet superradiance from mega-networks of tryptophan in biological architectures. *Journal of Physical Chemistry B* 128.17, pp. 4035–4046. DOI: 10.1021/acs.jpcb.3c07936.

Barrett, F. S. and R. R. Griffiths (2018). Classic hallucinogens and mystical experiences: Phenomenology and neural correlates. *Current Topics in Behavioral Neurosciences* 36, pp. 393–430. DOI: 10.1007/7854_2017_474.

Beggs, J. M. and D. Plenz (2003). Neuronal avalanches in neocortical circuits. *Journal of Neuroscience* 23.35, pp. 11167–11177. DOI: 10.1523/JNEUROSCI.23-35-11167.2003.

Bergson, H. (1896). *Matter and Memory.* Trans. by N. M. Paul and W. S. Palmer. Translation published 1991. Zone Books.

Birch, J., A. K. Schnell, and N. S. Clayton (2020). Dimensions of Animal Consciousness. *Trends in Cognitive Sciences* 24.10, pp. 789–801. DOI: 10.1016/j.tics.2020.07.007.

Blum, L. and M. Blum (2024). *AI Consciousness is Inevitable: A Theoretical Computer Science Perspective.* DOI: 10.48550/arXiv.2403.17101. arXiv: 2403.17101 [cs.AI].

Bowlby, J. (1969). *Attachment and Loss. Volume 1: Attachment.* Basic Books.

Brown, D. E. (1991). *Human Universals.* McGraw-Hill.

Butlin, P., R. Long, T. Bayne, Y. Bengio, J. Birch, D. Chalmers, A. Constant, et al. (2025). Identifying indicators of consciousness in AI systems. *Trends in Cognitive Sciences.* DOI: 10.1016/j.tics.2025.10.011.

Carhart-Harris, R. L. (2018). The entropic brain — revisited. *Neuropharmacology* 142, pp. 167–178. DOI: 10.1016/j.neuropharm.2018.03.010.

Carhart-Harris, R. L. and K. J. Friston (2019). REBUS and the anarchic brain. *Pharmacological Reviews* 71.3, pp. 316–344. DOI: 10.1124/pr.118.017160.

Carhart-Harris, R. L., R. Leech, P. J. Hellyer, M. Shanahan, A. Feilding, E. Tagliazucchi, D. R. Chialvo, and D. Nutt (2014). The entropic brain: A theory of conscious states informed by neuroimaging research with psychedelic drugs. *Frontiers in Human Neuroscience* 8, p. 20. DOI: 10.3389/fnhum.2014.00020.

Casali, A. G., O. Gosseries, M. Rosanova, M. Boly, S. Sarasso, K. R. Casali, S. Casarotto, et al. (2013). A theoretically based index of consciousness independent of sensory processing and behavior. *Science Translational Medicine* 5.198, 198ra105. DOI: 10.1126/scitranslmed.3006294.

Casarotto, S., A. Comanducci, M. Rosanova, S. Sarasso, M. Fecchio, M. Napolitani, A. Pigorini, et al. (2016). Stratification of Unresponsive Patients by an Independently Validated Index of Brain

Complexity. *Annals of Neurology* 80.5, pp. 718–729. DOI: 10.1002/ana.24779.

Chalmers, D. J. (1995a). "Absent qualia, fading qualia, dancing qualia." In: *Conscious Experience*. Ed. by T. Metzinger. Imprint Academic, pp. 309–328.

– (1995b). Facing up to the problem of consciousness. *Journal of Consciousness Studies* 2.3, pp. 200–219.

– (2010). "Phenomenal concepts and the explanatory gap." In: *The Character of Consciousness*. Oxford University Press, pp. 305–336. DOI: 10.1093/acprof:oso/9780195311105.003.0010.

– (Aug. 2023). Could a large language model be conscious? *Boston Review*. Edited from a 2022 NeurIPS talk. DOI: 10.48550/arXiv.2303.07103. arXiv: 2303.07103 [cs.AI].

Chittick, W. C. (1989). *The Sufi Path of Knowledge: Ibn al-Arabi's Metaphysics of Imagination*. Albany, NY: State University of New York Press.

Churchland, P. S. (1996). The hornswoggle problem. *Journal of Consciousness Studies* 3.5–6, pp. 402–408.

Cleeremans, A. (2008). "Consciousness: the radical plasticity thesis." In: *Models of Brain and Mind: Physical, Computational and Psychological Approaches*. Ed. by R. Banerjee and B. K. Chakrabarti. Vol. 168. Progress in Brain Research. Elsevier, pp. 19–33. DOI: 10.1016/S0079-6123(07)68003-0.

– (2011). The radical plasticity thesis: how the brain learns to be conscious. *Frontiers in Psychology* 2, p. 86. DOI: 10.3389/fpsyg.2011.00086.

Cocchi, L., L. L. Gollo, A. Zalesky, and M. Breakspear (2017). Criticality in the brain: A synthesis of neurobiology, models and cognition. *Progress in Neurobiology* 158, pp. 132–152. DOI: 10.1016/j.pneurobio.2017.07.002.

Cogitate Consortium and O. Ferrante (2025). Adversarial testing of global neuronal workspace and integrated information theories of consciousness. *Nature* 642.8066, pp. 133–142. DOI: 10.1038/s41586-025-08888-1.

Coleman, G. and T. Jinpa, eds. (2006). *The Tibetan Book of the Dead: The Great Liberation by Hearing in the Intermediate States.* Trans. by G. Dorje. London: Penguin Books.

Cowen, A. S., D. Keltner, F. Schroff, B. Jou, H. Adam, and G. Prasad (2021). Sixteen facial expressions occur in similar contexts worldwide. *Nature* 589, pp. 251–257. DOI: 10.1038/s41586-020-3037-7.

Damasio, A. (1999). *The Feeling of What Happens.* Harcourt.

– (2010). *Self Comes to Mind.* Pantheon.

Damasio, H. and A. Damasio (2024). Homeostatic Feelings and the Emergence of Consciousness. *Journal of Cognitive Neuroscience* 36.8, pp. 1653–1659. DOI: 10.1162/jocn_a_02119.

Dehaene, S. and J.-P. Changeux (2011). Experimental and theoretical approaches to conscious processing. *Neuron* 70.2, pp. 200–227. DOI: 10.1016/j.neuron.2011.03.018.

Dennett, D. C. (1991). *Consciousness Explained.* Little, Brown.

Deutsch, E. (1969). *Advaita Vedānta: A Philosophical Reconstruction.* Honolulu: East-West Center Press.

Dung, L. and A. Newen (2023). Profiles of Animal Consciousness: A Species-Sensitive, Multidimensional Approach. *Cognition* 235, p. 105409. DOI: 10.1016/j.cognition.2023.105409.

Edelman, G. M. (1989). *The Remembered Present.* Basic Books.

Edelman, G. M. and G. Tononi (2000). *A Universe of Consciousness.* Basic Books.

Esteban, F. J., J. A. Galadí, J. A. Langa, J. R. Portillo, and F. Soler-Toscano (2018). Informational structures: A dynamical system approach for integrated information. *PLOS Computational Biology* 14.9, e1006154. DOI: 10.1371/journal.pcbi.1006154.

Everitt, B. J. and T. W. Robbins (2016). Drug addiction: Updating actions to habits to compulsions ten years on. *Annual Review of Psychology* 67, pp. 23–50. DOI: 10.1146/annurev-psych-122414-033457.

Fekete, T. and S. Edelman (2011). Towards a computational theory of experience. *Consciousness and Cognition* 20.3, pp. 807–827. DOI: 10.1016/j.concog.2011.02.010.

– (2012). "The (lack of) mental life of some machines." In: *Being in Time: Dynamical Models of Phenomenal Experience.* Ed. by S. Edelman, T. Fekete, and N. Zach. John Benjamins, pp. 95–120.

Ferrarelli, F., M. Massimini, S. Sarasso, A. Casali, B. A. Riedner, G. Angelini, G. Tononi, and R. A. Pearce (2010). Breakdown in Cortical Effective Connectivity during Midazolam-Induced Loss of Consciousness. *Proceedings of the National Academy of Sciences* 107.6, pp. 2681–2686. DOI: 10.1073/pnas.0913008107.

Fraser-Taliente, K., S. Kantamneni, E. Ong, D. Mossing, C. Lu, P. C. Bogdan, E. Ameisen, et al. (May 2026). *Natural Language Autoencoders Produce Unsupervised Explanations of LLM Activations.* Transformer Circuits Thread. URL: https://transformer-circuits.pub/2026/nla/index.html.

Freeman, W. J. (1999). *How Brains Make Up Their Minds.* Weidenfeld & Nicolson.

Friston, K. (2010). The free-energy principle: A unified brain theory? *Nature Reviews Neuroscience* 11.2, pp. 127–138. DOI: 10.1038/nrn2787.

– (2019). *A free energy principle for a particular physics.* Tech. rep. 1906.10184. arXiv. DOI: 10.48550/arXiv.1906.10184.

Friston, K., L. Da Costa, N. Sajid, C. Heins, K. Ueltzhöffer, G. A. Pavliotis, and T. Parr (2023). The free energy principle made simpler but not too simple. *Physics Reports* 1024, pp. 1–29. DOI: 10.1016/j.physrep.2023.07.001.

Friston, K., C. Thornton, and A. Clark (2012). Free-energy minimization and the dark-room problem. *Frontiers in Psychology* 3, p. 130. DOI: 10.3389/fpsyg.2012.00130.

Garfield, J. L. (1995). *The Fundamental Wisdom of the Middle Way: Nagarjuna's Mulamadhyamakakarika.* New York: Oxford University Press. DOI: 10.1093/oso/9780195103175.001.0001.

Garland, J., E. Bradley, and J. D. Meiss (2016). Exploring the topology of dynamical reconstructions. *Physica D* 334, pp. 49–59. DOI: 10.1016/j.physd.2016.03.006.

Goff, P. (2019). *Galileo's Error: Foundations for a New Science of Consciousness.* Pantheon Books.

Griffiths, R. R., M. W. Johnson, W. A. Richards, B. D. Richards, U. McCann, and R. Jesse (2011). Psilocybin occasioned mystical-type experiences: Immediate and persisting dose-related effects. *Psychopharmacology* 218.4, pp. 649–665. DOI: 10.1007/s00213-011-2358-5.

Griffiths, R. R., W. A. Richards, U. McCann, and R. Jesse (2006). Psilocybin can occasion mystical-type experiences having substantial and sustained personal meaning and spiritual significance. *Psychopharmacology* 187.3, pp. 268–283. DOI: 10.1007/s00213-006-0457-5.

Hameroff, S. and R. Penrose (2014). Consciousness in the universe: A review of the 'Orch OR' theory. *Physics of Life Reviews* 11.1, pp. 39–78. DOI: 10.1016/j.plrev.2013.08.002.

Heil, M. and R. Karban (2010). Explaining evolution of plant communication by airborne signals. *Trends in Ecology & Evolution* 25.3, pp. 137–144. DOI: 10.1016/j.tree.2009.09.010.

Henrich, J., S. J. Heine, and A. Norenzayan (2010). The weirdest people in the world? *Behavioral and Brain Sciences* 33.2–3, pp. 61–83. DOI: 10.1017/S0140525X0999152X.

Hinwar, R. P. and A. J. Lambert (2021). Anauralia: the silent mind and its association with aphantasia. *Frontiers in Psychology* 12, p. 744213. DOI: 10.3389/fpsyg.2021.744213.

Hoffman, D. D. (2019). *The Case Against Reality.* Norton.

Hoffman, D. D. and C. Prakash (2014). Objects of consciousness. *Frontiers in Psychology* 5, p. 577. DOI: 10.3389/fpsyg.2014.00577.

Hofstadter, D. R. (1979). *Gödel, Escher, Bach.* Basic Books.

– (2007). *I Am a Strange Loop.* Basic Books.

Huxley, A. (1954). *The Doors of Perception.* Chatto & Windus.

Jackson, F. (1982). Epiphenomenal qualia. *Philosophical Quarterly* 32.127, pp. 127–136. DOI: 10.2307/2960077.

Jackson, F. (2003). Mind and Illusion. *Royal Institute of Philosophy Supplements* 53, pp. 251–271. DOI: 10.1017/S1358246100008365.

Jackson, J. C., J. Watts, T. R. Henry, J.-M. List, R. Forkel, P. J. Mucha, S. J. Greenhill, R. D. Gray, and K. A. Lindquist (2019). Emotion semantics show both cultural variation and universal structure. *Science* 366.6472, pp. 1517–1522. DOI: 10.1126/science.aaw8160.

James, W. (1898). *Human immortality: Two supposed objections.* Ingersoll Lecture.

John of the Cross (1991). *The Collected Works of St. John of the Cross.* Trans. by K. Kavanaugh and O. Rodriguez. Revised edition. Washington, DC: ICS Publications.

Johnson, S. E. (2026). *Engagement, openness, and consolidation: A three-variable dynamical system for psychological transition.* Zenodo. DOI: 10.5281/zenodo.19422286.

Johnson, S. E., W. J. Song, E. K. Anderson, C. C. Gerbi, S. S. Vel, D. J. Prior, and M. Stipp (2025). EBSD-based calibration of differential stress from experimentally deformed Black Hills quartzite using the perimeter-area fractal dimension. *Journal of Geophysical Research: Solid Earth* 130, e2024JB030866. DOI: 10.1029/2024JB030866.

Jung, C. G. (1959). *The Archetypes and the Collective Unconscious.* Ed. by H. Read, M. Fordham, and G. Adler. Trans. by R. F. C. Hull. Vol. 9, Part 1. The Collected Works of C. G. Jung. London: Routledge & Kegan Paul.

Karst, J., M. D. Jones, and J. D. Hoeksema (2023). Positive citation bias and overinterpreted results lead to misinformation on common mycorrhizal networks in forests. *Nature Ecology & Evolution* 7.4, pp. 501–511. DOI: 10.1038/s41559-023-01986-1.

Kelly, J. F., K. Humphreys, and M. Ferri (2020). Alcoholics Anonymous and other 12-step programs for alcohol use disorder. *Cochrane Database of Systematic Reviews* 3, p. CD012880. DOI: 10.1002/14651858.CD012880.pub2.

Kelso, J. A. S. (1995). *Dynamic Patterns.* MIT Press.

Keltner, D. and J. Haidt (2003). Approaching awe, a moral, spiritual, and aesthetic emotion. *Cognition and Emotion* 17.2, pp. 297–314. DOI: 10.1080/02699930302297.

Lafer-Sousa, R., K. L. Hermann, and B. R. Conway (2015). Striking Individual Differences in Color Perception Uncovered by The Dress Photograph. *Current Biology* 25.13, R545–R546. DOI: 10.1016/j.cub.2015.04.053.

Leary, T., R. Metzner, and R. Alpert (1964). *The Psychedelic Experience: A Manual Based on the Tibetan Book of the Dead.* University Books.

Levine, J. (1983). Materialism and qualia: The explanatory gap. *Pacific Philosophical Quarterly* 64.4, pp. 354–361. DOI: 10.1111/j.1468-0114.1983.tb00207.x.

Li, Q. (2010). Effect of forest bathing trips on human immune function. *Environmental Health and Preventive Medicine* 15.1, pp. 9–17. DOI: 10.1007/s12199-008-0068-3.

Li, Q., K. Morimoto, A. Nakadai, H. Inagaki, M. Katsumata, T. Shimizu, and T. Kawada (2007). Forest bathing enhances human natural killer activity and expression of anti-cancer proteins. *International Journal of Immunopathology and Pharmacology* 20.2, Suppl 2, pp. 3–8. DOI: 10.1177/03946320070200S202.

Lindsey, J. (2025). *Emergent introspective awareness in large language models.* Transformer Circuits Thread. URL: https://transformer-circuits.pub/2025/introspection/index.html.

Long, R., J. Sebo, P. Butlin, K. Finlinson, K. Fish, J. Harding, J. Pfau, T. Sims, J. Birch, and D. Chalmers (Nov. 2024). *Taking AI Welfare Seriously.* arXiv: 2411.00986. URL: https://arxiv.org/abs/2411.00986.

Lorenz, E. N. (1963). Deterministic nonperiodic flow. *Journal of the Atmospheric Sciences* 20.2, pp. 130–141. DOI: 10.1175/1520-0469(1963)020<0130:DNF>2.0.CO;2.

Lucas, J. R. (1961). Minds, machines and Gödel. *Philosophy* 36.137, pp. 112–127. DOI: 10.1017/S0031819100057983.

Massimini, M., F. Ferrarelli, R. Huber, S. K. Esser, H. Singh, and G. Tononi (2005). Breakdown of cortical effective connectivity during sleep. *Science* 309.5744, pp. 2228–2232. DOI: 10.1126/science.1117256.

Massimini, M., F. Ferrarelli, M. J. Murphy, R. Huber, B. A. Riedner, S. Casarotto, and G. Tononi (2010). Cortical Reactivity and Effective Connectivity during REM Sleep in Humans. *Cognitive Neuroscience* 1.3, pp. 176–183. DOI: 10.1080/17588921003731578.

McGinn, B. (2001). *The Mystical Thought of Meister Eckhart: The Man from Whom God Hid Nothing.* New York: Crossroad.

Metzinger, T. (2003). *Being No One.* MIT Press.

Milton, A. L. (2023). Drug memory reconsolidation: From molecular mechanisms to the clinical context. *Translational Psychiatry* 13, p. 370. DOI: 10.1038/s41398-023-02666-1.

Moos, R. H. and B. S. Moos (2006). Rates and predictors of relapse after natural and treated remission from alcohol use disorders. *Addiction* 101.2, pp. 212–222. DOI: 10.1111/j.1360-0443.2006.01310.x.

Moyal, R., T. Fekete, and S. Edelman (2020). Dynamical emergence theory (DET): A computational account of phenomenal consciousness. *Minds and Machines* 30.1, pp. 1–21. DOI: 10.1007/s11023-020-09516-9.

Nagel, T. (1974). What is it like to be a bat? *Philosophical Review* 83.4, pp. 435–450. DOI: 10.2307/2183914.

Ninkovic, V., D. Markovic, and M. Rensing (2021). Plant volatiles as cues and signals in plant communication. *Plant, Cell & Environment* 44.4, pp. 1030–1043. DOI: 10.1111/pce.13910.

Oizumi, M., L. Albantakis, and G. Tononi (2014). From the phenomenology to the mechanisms of consciousness: Integrated Information Theory 3.0. *PLOS Computational Biology* 10.5, e1003588. DOI: 10.1371/journal.pcbi.1003588.

Parr, T., L. Da Costa, and K. Friston (2020). Markov blankets, information geometry and stochastic thermodynamics. *Philosophical Transactions of the Royal Society A* 378.2164, p. 20190159. DOI: 10.1098/rsta.2019.0159.

Parr, T., G. Pezzulo, and K. J. Friston (2022). *Active Inference: The Free Energy Principle in Mind, Brain, and Behavior.* Cambridge, MA: MIT Press.

Penrose, R. (1989). *The Emperor's New Mind: Concerning Computers, Minds, and the Laws of Physics.* Oxford: Oxford University Press.

– (1994). *Shadows of the Mind: A Search for the Missing Science of Consciousness.* Oxford: Oxford University Press.

Popper, K. R. (1959). *The Logic of Scientific Discovery.* London: Hutchinson.

Prakash, C., K. D. Stephens, D. D. Hoffman, M. Singh, and C. Fields (2021). Fitness beats truth in the evolution of perception. *Acta Biotheoretica* 69.3, pp. 319–341. DOI: 10.1007/s10441-020-09400-0.

Rankaduwa, S. and A. M. Owen (2023). Psychedelics, entropic brain theory, and the taxonomy of conscious states: A summary of debates and perspectives. *Neuroscience of Consciousness* 2023.1, niad001. DOI: 10.1093/nc/niad001.

Robertson, C. E. and S. Baron-Cohen (2017). Sensory perception in autism. *Nature Reviews Neuroscience* 18, pp. 671–684. DOI: 10.1038/nrn.2017.112.

Rosenthal, D. M. (2005). *Consciousness and Mind.* Oxford University Press.

Rossi, E. L. (1989). Archetypes as strange attractors. *Psychological Perspectives* 20.1, pp. 4–15.

Russell, B. (1927). *The Analysis of Matter.* Kegan Paul.

Schartner, M. M., R. L. Carhart-Harris, A. B. Barrett, A. K. Seth, and S. D. Muthukumaraswamy (2017). Increased spontaneous MEG signal diversity for psychoactive doses of ketamine, LSD and psilocybin. *Scientific Reports* 7, p. 46421. DOI: 10.1038/srep46421.

Scholem, G. G. (1954). *Major Trends in Jewish Mysticism.* 3rd ed. New York: Schocken Books.

Seth, A. K. (2021). *Being You: A New Science of Consciousness.* New York: Dutton.

– (Apr. 21, 2025). Conscious Artificial Intelligence and Biological Naturalism. *Behavioral and Brain Sciences.* DOI: 10.1017/S0140525X25000032.

Seth, A. K. and M. Tsakiris (2018). Being a Beast Machine: The Somatic Basis of Selfhood. *Trends in Cognitive Sciences* 22.11, pp. 969–981. DOI: 10.1016/j.tics.2018.08.008.

Shinozuka, K., K. Jerotic, P. Mediano, A. T. Zhao, K. H. Preller, R. Carhart-Harris, and M. L. Kringelbach (2024). Synergistic, multi-level understanding of psychedelics: Three systematic reviews and meta-analyses of their pharmacology, neuroimaging and phenomenology. *Translational Psychiatry* 14, p. 485. DOI: 10.1038/s41398-024-03187-1.

Simard, S. W., K. J. Beiler, M. A. Bingham, J. R. Deslippe, L. J. Philip, and F. P. Teste (2012). Mycorrhizal networks: Mechanisms, ecology and modelling. *Fungal Biology Reviews* 26.1, pp. 39–60. DOI: 10.1016/j.fbr.2012.01.001.

Song, Y., J. Sohl-Dickstein, D. P. Kingma, A. Kumar, S. Ermon, and B. Poole (2021). "Score-Based Generative Modeling through Stochastic Differential Equations." In: *International Conference on Learning Representations (ICLR).*

Sparrow, C. (1982). *The Lorenz Equations: Bifurcations, Chaos, and Strange Attractors.* Springer-Verlag.

Strawson, G. (2006). Realistic monism. *Journal of Consciousness Studies* 13.10-11, pp. 3–31.

Sugden, S. G., G. Merlo, and S. Manger (2024). Strengthening neuroplasticity in substance use recovery through lifestyle intervention.

American Journal of Lifestyle Medicine 18.5, pp. 648–656. DOI: 10.1177/15598276241242016.

Tagliazucchi, E., R. L. Carhart-Harris, R. Leech, D. Nutt, and D. R. Chialvo (2014). Enhanced repertoire of brain dynamical states during the psychedelic experience. *Human Brain Mapping* 35.11, pp. 5442–5456. DOI: 10.1002/hbm.22562.

Taiz, L., D. Alkon, A. Draguhn, A. Murphy, M. Blatt, C. Hawes, G. Thiel, and D. G. Robinson (2019). Plants neither possess nor require consciousness. *Trends in Plant Science* 24.8, pp. 677–687. DOI: 10.1016/j.tplants.2019.05.008.

Tegmark, M. (2000). Importance of quantum decoherence in brain processes. *Physical Review E* 61.4, pp. 4194–4206. DOI: 10.1103/PhysRevE.61.4194.

The New York Declaration on Animal Consciousness (2024). Public declaration, April 19, 2024. URL: https://sites.google.com/nyu.edu/nydeclaration/declaration.

Thompson, E. (2007). *Mind in Life.* Harvard University Press.

Tononi, G. (2004). An information integration theory of consciousness. *BMC Neuroscience* 5, p. 42. DOI: 10.1186/1471-2202-5-42.

– (2008). Consciousness as integrated information: A provisional manifesto. *Biological Bulletin* 215.3, pp. 216–242. DOI: 10.2307/25470707.

Upaniṣads (1996). Trans. by P. Olivelle. Oxford World's Classics. Oxford: Oxford University Press.

Van Eenwyk, J. R. (1991). Archetypes: The strange attractors of the psyche. *Journal of Analytical Psychology* 36.1, pp. 1–25. DOI: 10.1111/j.1465-5922.1991.00001.x.

– (1997). *Archetypes and Strange Attractors.* Inner City Books.

Varela, F. J., E. Thompson, and E. Rosch (1991). *The Embodied Mind.* MIT Press.

Volkow, N. D., G. F. Koob, and A. T. McLellan (2016). Neurobiologic advances from the brain disease model of addiction. *New England Journal of Medicine* 374.4, pp. 363–371. DOI: 10.1056/NEJMra1511480.

Ward, J. (2013). Synesthesia. *Annual Review of Psychology* 64, pp. 49–75. DOI: 10.1146/annurev-psych-113011-143840.

Ward, J., C. Hoadley, J. E. A. Hughes, P. Smith, C. Allison, S. Baron-Cohen, and J. Simner (2017). Atypical sensory sensitivity as a shared feature between synaesthesia and autism. *Scientific Reports* 7, p. 41155. DOI: 10.1038/srep41155.

Webb, T. W. and M. S. A. Graziano (2015). The attention schema theory. *Frontiers in Psychology* 6, p. 500. DOI: 10.3389/fpsyg.2015.00500.

Wilhelm, R. (1931). *The Secret of the Golden Flower: A Chinese Book of Life.* Trans. by C. F. Baynes. With a commentary by C. G. Jung. Kegan Paul, Trench, Trubner & Co.

Wissner-Gross, A. D. and C. E. Freer (2013). Causal entropic forces. *Physical Review Letters* 110.16, p. 168702. DOI: 10.1103/PhysRevLett.110.168702.

Zeman, A., M. Dewar, and S. Della Sala (2015). Lives without Imagery: Congenital Aphantasia. *Cortex* 73, pp. 378–380. DOI: 10.1016/j.cortex.2015.05.019.

About the Author

Scott E. Johnson is a professor of Earth and climate sciences at the University of Maine, where his research focuses on the deformation of the solid Earth. His broad interest is in the physics of flow, whether in water, ice, atmospheres, or rocks, and how these systems respond to evolving forces. He also studies the inner landscape through depth psychology, contemplative practice, and direct experience.

The Shape of Consciousness grew from the convergence of those two paths and the recognition that the dynamical principles governing how physical systems organize may also describe how finite minds organize into experience.

Scott lives in Maine. The companion site for this book is located at innerattractors.com. His work appears in the peer-reviewed scientific literature and at Inner Exploration Labs (innerexplore.com).

www.ingramcontent.com/pod-product-compliance
Lightning Source LLC
LaVergne TN
LVHW022156200926
848481LV00006B/1768
* 9 7 9 8 9 9 6 5 4 5 1 0 0 *